Sustainability and Innov

Coordinating Editor

Jens Horbach
University of Applied Sciences Anhalt, Bernburg, Germany

Series Editors

Eberhard Feess
RWTH Aachen, Germany

Jens Hemmelskamp
University of Heidelberg, Germany

Joseph Huber
University of Halle-Wittenberg, Germany

René Kemp
University of Maastricht, The Netherlands

Marco Lehmann-Waffenschmidt
Dresden University of Technology, Germany

Arthur P. J. Mol
Wageningen Agricultural University, The Netherlands

Sustainability and Innovation

Published Volumes:

Jens Horbach (Ed.)
Indicator Systems for Sustainable Innovation
2005. ISBN 3-7908-1553-5

Bernd Wagner, Stefan Enzler (Eds.)
Material Flow Management
2006. ISBN 3-7908-1591-8

Andreas Ahrens · Angelika Braun
Arnim von Gleich · Kerstin Heitmann
Lothar Lißner

Hazardous Chemicals in Products and Processes

Substitution as an Innovative Process

With Contributions of
Andrea Effinger, Matthias Weiß, Claudia Wölk

With 21 Figures

Physica-Verlag

A Springer Company

Andreas Ahrens
Ökopol
Nernstweg 32–34
22765 Hamburg
Germany
ahrens@oekopol.de

Kerstin Heitmann
Ökopol
Nernstweg 32–34
22765 Hamburg
Germany
heitmann@oekopol.de

Angelika Braun
Kooperationsstelle Hamburg
Besenbinderhof 60
20097 Hamburg
Germany
angelikabraun-hh@web.de

Lothar Lißner
Kooperationsstelle Hamburg
Besenbinderhof 60
20097 Hamburg
Germany
lissner_koop@public.uni-hamburg.de

Professor Arnim von Gleich
Universität Bremen
FB 04 Produktionstechnik
Postfach 330440
28334 Bremen
Germany
gleich@uni-bremen.de

This project was sponsored by the Federal Ministry of Education and Research under the funding program "Frameworks for Innovation towards Sustainability (RIW)" (project number 07RIW) – the authors assume sole responsibility for the contents of this report.

ISSN 1860-1030
ISBN-10   3-7908-1642-6 Physica-Verlag Heidelberg New York
ISBN-13   978-3-7908-1642-6 Physica-Verlag Heidelberg New York

Physica is a part of Springer Science+Business Media

springeronline.com

© Physica-Verlag Heidelberg 2006
Printed in Germany

Cover-Design: Erich Kirchner, Heidelberg

SPIN 11544319   88/3153-5 4 3 2 1 0 – Printed on acid-free paper

Contents

Introduction

Damage and risks caused by mineral fibres, metal dust and organic chemicals all occur in the course of industrial history: lung cancer as the result of inhaling asbestos fibres; brain damage and cancer due to solvent vapours at the workplace and also in chemical cleaning processes; global dispersion and bioaccumulation of pesticides such as DDT and of industrial chemicals such as PCBs; of heavy metals such as mercury, lead and cadmium; destruction of the ozone layer due to CFCs as well as potential risks related to industrial chemicals with hormone-like effects (e. g. phthalates, TBT[1]) and the presence of potential carcinogenic acryl amide in French fries.

From the perspective of sustainability, the development of such problems is particularly relevant, as they are rather insidious and not immediately apparent, and may be detected too late and thus no longer be amenable to remedy. In the field of hazardous substances, this applies to those substances that are persistent (i.e. are not or are only very slowly biologically or photo chemically degradable), are mobile (gaseous, dust-like, soluble in water), and bio-accumulative and can thus disperse globally and/or accumulate in the biological food chains. Also substances that, even after a long delay, trigger serious health risks in the case of chronic exposure to small doses (e.g. cancer, diminished reproduction capabilities) should be mentioned here.

The history of endeavours to reduce such risks is just as lengthy as the history of substance-related risks. In doing so, state regulative practices initially concentrated on occupational health and safety, and later also on media-related environmental protection. It was not until chemicals legislation was introduced at the beginning of the 1980s that regulations also directly related to the production and marketing of hazardous substances. In all three areas of regulation, occupational health and safety[2], environmental protection[3] and chemicals regulation[4], replace-

---

[1] PCB: polychlorinated biphenyls, TBT: tributyl tin, CFC: chlorofluorocarbons

[2] Cf. e.g. the substitution principle in the German Ordinance on Hazardous Substances (§16 and §36). Please refer to 'Technical Rules for Hazardous Substances´ TRGS 440 for the recommended procedure of substitution.

[3] Certain regulations contained in the air legislation and also in the annexes to the Waste Water Ordinance include the substitution of hazardous substances as the best available technology and thus attempt to put into operation the precautionary principle (cf. UBA texts 88/99, Guidance Manual for Formulators and Other Professional Users of Chemicals).

ment/substitution of hazardous substances by less dangerous substances is emphasised as a key element of risk management. Companies, driven not only by state regulation but also by well understood self-interest, were repeatedly faced with the question as to whether less dangerous substances could be used in individual applications.

In actual fact a great deal has also been undertaken in this area in the past. Many of the hazardous substances mentioned above have now more or less disappeared from the market. Some were banned (e.g. DDT, CFCs, PCBs), their functions now being performed by less hazardous substances. Other hazardous substances have at least been considerably curtailed in their use with safety requirements being imposed (e.g. chlorinated solvents, highly toxic heavy metals). This has also led to the reduction of risks emanating from hazardous substances in many areas.

The history of the substitution of hazardous chemicals could be considered a success story. If it is examined more closely, however, a range of as yet unresolved tasks are still evident (cf. chapter 6). This basically concerns two problem areas: the fundamental ability and willingness to substitute hazardous substances and the question whether the substitute is actually any less dangerous.

Considerable inertia in established practices can be observed everywhere, a tediousness and resistance to change, against which the substitution of hazardous substances has to struggle. Although in many cases the problems of hazardous substances are evident to a large extent, and although substitutes are available, indeed available for many areas of application, the substitution process is not progressing. The substitution of asbestos was a typically extremely tedious process (cf. chapter 2.2).

If we consider the process of hazardous substance substitution as an innovative process, what we are dealing with here is, firstly, a problem of a lack of willingness to innovate or a lack of the ability to be innovative. Secondly, the uncertainty surrounding the direction of innovation also plays a major role. Is the substitute substance in fact any less dangerous or does it entail new possibly as yet unknown dangers and problems – as was the case with the introduction of CFCs as a predicted low-risk substitute for ammonia as a refrigerant –?

The research project 'Options for viable innovation systems for successful substitution of hazardous substances' (or SubChem, for short)[5] was concerned with the problem of the ability to be innovative and of the direction of innovation with regard to risk reduction by hazardous substance substitution as part of the programme 'Framework conditions for innovation towards sustainability' funded by

---

[4] The VOC Directive (EU Directive 1999/13/EC on the limitation of emissions of volatile organic compounds due to the use of organic solvents in certain activities and installations) thus contains an obligation to substitute hazardous substances.

[5] FKZ 07RIW4, for further information cf. http://www.subchem.de. Information on the research program can be consulted at: http://www.riw-netzwerk.de

the Federal Ministry for Education and Research (BMBF)'[6]. The objective of this research project was to discover - on the basis of 13 case studies - under which framework conditions and in which constellations of actors the substitution of hazardous substances is encouraged or is impeded. In particular, this took into account the specific regulatory systems, the conditions on the markets as well as the ongoing public debate, in the course of which a series of substances hit the headlines as the so-called 'contaminant of the month'.

---

[6] The project partners were the Hamburg University of Applied Sciences (HAW), Ökopol – Institute for Environmental Strategies GmbH Hamburg and the Co-operation Office of the German Trade Unions Ass./Universities of Hamburg

1 Summary of most important results

By way of introduction, the most important project results are outlined here. Initially, the project focuses on the main issues, i.e. the ability to be innovative and the direction of innovation, and also on the current reforms in chemicals legislation taking place at EU level. After that, selected results are highlighted and explained.

1.1 Aspects of the ability to be innovative

The attempt to understand the frequently faced tediousness of substitution processes is firstly directed at individual actors[7], their motives and their opportunities for influence, and also at the way they utilise these opportunities or, rather, do not utilise them. The 'roles' of these actors can then immediately be sub-divided into promoters and blockers, and an attempt will then be made to explain the success or failure of substitution as a consequence of a certain distribution of interests and powers. In fact, it always comes down to people who promote or block innovations. This begins with the entrepreneurial personality as illustrated by Schumpeter, who performs its work of 'creative destruction', via 'entrepreneurs' who are said to be indifferent to occupational health and safety, consumer protection or environmental protection, to cultural pessimists and 'luddites', who always aimed to impede one technology or another.

With regard to innovation processes the significance and the direct effectiveness of individual committed promoters or blockers with their individual motives such as profit or occupational health and safety, consumer protection or environmental protection should, however, not be overestimated. Although committed promoters do play an important part in most substitution processes, a closer look at the individual case samples very quickly reveals their structural futility. Complex innovative processes cannot be moved by a limited number of actors or even by individuals alone. Many substitution processes simply do not progress, despite the fact that we cannot observe any definite 'opponents'. These innovations are not impeded, they only become stuck, as the 'inertia of the system' is simply too high. In order to gain an appropriate comprehension of the ability (and not just the

---

[7] According to the German word `Akteure´, `actors´ in an innovation system means manufactures, importers and users of chemicals (actors in the supply chain, economic actors) as well as authorities, science, public interest groups and other participants outside the supply chain (cf. Figure 1).

willingness) for hazardous substance substitution (the ability to be innovative), we therefore must not solely look at individual participants in a supply chain, their interests and opportunities for influence. It is more important to have an overall view of the in some cases highly complex 'constellations of actors', including the 'framework conditions', which have an either encouraging or preventive effect on substitution processes as legislation, competitive conditions and public debates (cf. Figure 1). This is the reason why a system-theories approach was chosen in the SubChem project and the concept and heuristic model of 'innovation systems' was used. Overcoming the pure actors' perspective may contain in itself the risk of losing the action relevance of the expected results. Nevertheless a differentiated systems view should improve retroactively the individual actors' opportunities for action considerably. If the participants are able to develop a differentiated 'system comprehension', they can also better exploit their (albeit limited) opportunities for influence.

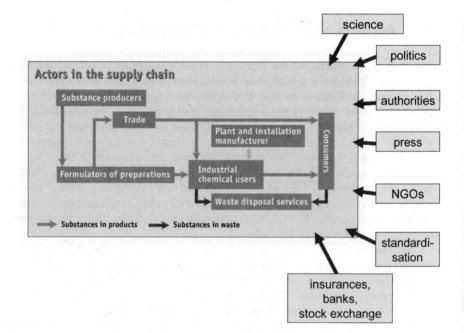

Figure 1. Actors in the innovation system: inside and outside the supply chain

1.2 Aspects of direction of innovation

Misguided substitutions such as the introduction of CFCs as a 'safe' substitute for ammonia as a refrigerant were already mentioned in the introduction. The example of CFCs may be an extreme case. But also the case studies examined in the project such as 'the substitution of asbestos by non-bio soluble mineral fibres in construction uses' and also 'the replacement of inflammable hydrocarbons by non-inflammable chlorinated hydrocarbons in metal cleaning' are examples of the existing major problems in orientation with regard to the direction of innovation. These uncertainties slow down many substitution processes additionally. Many substitution processes are also not developed precisely as there is widespread uncertainty as to whether the substitute really does help reduce risks. The latter is certainly so in the case of plasticisers in PVC products, where incomplete toxicological knowledge was quoted as an argument against substituting DEHP by citrate esters in sensitive applications, and at least partly in the case of resistance against the criterion "bio solubility" in the substitution of mineral fibres. The realisation from the case studies that the opponents of an innovation are too prone to utilise the (fatal) argument of insufficient knowledge in case of conflict also demonstrates only the tip of the problem from the position of the participants. The endless passing on of non-realisable 'risk or non-dangerous burdens of proof' is doubtless a game that is as futile as it is widespread. The industry's lack of responsibility can be denounced or troublesome demands for substitutions can also be blocked ad infinitum. Nevertheless, the lack of certainty in orientation in the substitution of hazardous substances is not a motivation problem, but rather it is the problem of dealing with lack of knowledge, which can only be resolved structurally, in an appropriate way.

To be precise in many cases we know just as little about the problematic side effects and consequential effects of substances that have already been employed as we do about the alternatives. In practice, however, comparable uncertainties appear to have completely different effects. They generally have a greater effect against substitution and/or the substitute[8]. Especially in the case of planned changes in common (and possibly highly problematic) practices to date, it is easy to highlight the many uncertainties related to innovation. The innovator is generally faced with the obligation to demonstrate a greater 'burden of evidence' than parties wishing to leave things as they are. The uncertainties have the structural effect of discouraging innovation, even if they are not especially 'exploited' by participants[9].

However, the problem of an appropriate way of dealing with incomplete knowledge and major uncertainties is not only present in the case of hazardous substance substitution. This is a fundamental problem for any innovation. In this

---

[8] Current European chemicals legislation even amplifies this problem. Substances marketed in Europe after 1981 (or new substances, as they are called) are subject to high demands in respect of (eco)toxicological chemicals testing, while those substances already existing on the market at this time (or existing substances, as they are called) may continue to be used without any testing. This differentiation is no longer applicable in the run-up to the new EU chemicals policy (REACH).

[9] If certain substances are brought into 'disrepute' amid great public attention, the reverse effect is however sometimes also observed. Based on the principle of 'anything but this substance', any alternative may gain a structural advantage.

respect innovation and risk are inseparably interlinked. Substitution of hazardous substances is not fundamentally different from other forms of innovation. Companies had to learn to deal rationally with the economic risks, which is also just as applicable when dealing with technical, health and ecological risks. Knowledge about risks and hazards is restricted and incomplete in all these areas. Thus ways of dealing appropriately with the remaining uncertainties have to be developed.

In addition to the general systems inertia already mentioned, the uncertainties that always remain in view of all innovations represent significantly 'more effective' barriers to innovation than all positive or negative motives and interests of the participants, which could be ascertained in the various case studies. The need to focus on innovation systems at supply chain level as well as overcoming system inertia and uncertainties and/or lack of knowledge, as fundamental barriers to innovation are some of the important findings of the SubChem project[10]. However, this in turn produces new questions.

1. What does flexibility and/or inertia of innovation systems depend on (the ability to be innovative)?
2. What opportunities exist, despite the remaining major uncertainties, to promote innovations and/or the substitution of hazardous substances successfully and in the appropriate direction (dealing with lack of knowledge, decision on direction of innovation)?

1.3   Model of "innovation systems at supply chain level"

Comparatively early on in the research process a heuristic model of 'innovation systems on supply chain level of hazardous substance substitution' was developed in co-operation with the two other 'chemicals projects'[11] in the research programme. In this model, the supply chain forms the central point and four main influencing factors affect the system: the regulative pull of application-related legislation in the area of occupational health and safety; environmental protection and consumer protection; the regulative push of chemicals legislation (regulations governing market entry); the pull of market demand and the push of scientific/technical developments, which continually present new solution options (cf. Figure 6, Chapter 3.3.2).

The SubChem research process took place in a constant process of interaction between a deeper analysis of the system and understanding of the system and the empirical studies on cases. The cases are neither self-explanatory nor is innovation research already so far advanced that hypotheses derived from an established and

---

[10] These findings are not necessarily surprising. They tie in well with the everyday experience of all 'innovators', all 'entrepreneurs', who really wish to carry out an entrepreneurial activity, but also all political reformers, revolutionaries etc.

[11] Within the :[riw]-framework three research projects were engaged in issues related to chemicals legislation: SubChem, COIN and INNOCHEM (cf. http://www.riw-netzwerk.de/projekte/riw_00_02_00.htm)

recognised set of theories would only require 'empirical' verification. In the case studies theoretically based hypotheses from the system view could be examined empirically in descending abstraction (i.e. deductively) and, conversely, hypotheses about the significance of certain constellations of framework conditions, actors and their opportunities for influence were able to be generalised from the case studies in ascending abstraction (i.e. inductively). Both abstraction directions can only be differentiated as an ideal type; they are always interlinked and they have also been passed through iteratively several times in the course of the research process. In this way, with regard to the subject "the ability to be innovative", two types of results were developed: on the one hand an improved 'systemic' perception including a differentiated model of the innovation system and, on the other hand, a set of hypotheses about the effects (and/or possibilities of effects) of certain framework conditions and actors.

1.4   Current developments in chemicals regulation

If the current regulative framework conditions in the area of hazardous substances are examined critically under the aspects of both health and environmental protection and also with regard to the effects on the ability to be innovative and on the direction of innovation, the conclusion may be reached that the current reform of EU chemicals legislation in accordance with REACH[12] is indeed a step in the right direction. The predominant regulation of chemicals' application conditions, initially from a historical aspect, (i.e. regulation pull in our model of innovation system) suffers greatly from the diversity of the specific situations, with the result that there is an excessive deficit in implementation. In this respect, it is entirely logical to create a more regulative approach for the marketing of chemicals (i.e. regulation push in our model of innovation system) and from this approach also to move on to the application conditions (or the various exposure scenarios, respectively). REACH also compensates for some of the current serious structural disadvantages of new substances (and/or the trend to prefer existing substances, which curbs innovation). The (risk) communication along the supply chain promoted by REACH lastly supports the long-overdue re-orientation of innovation systems, which are still too branch-specific. The chemical industry had always seen itself as a substance manufacturer and had organised itself accordingly. The important innovations, which were the reason for the strong competitive position of the German chemical industry up to the 1960s and 1970s, were in fact essentially developed in laboratories[13]. Frequently at that time a new interesting substance was first synthesised, after which the search for possible lucrative areas of application was

---

[12] REACH is an acronym comprising the most important elements of the new chemicals legislation at EU level: *Registration, Evaluation and Authorisation of CHemicals*, cf. http://europa.eu.int/comm/enterprise/chemicals/chempol/whitepaper/reach.htm

[13] Cf. Grupp et al 2002, Dominguez-Lacasa et al 2003

launched[14]. The innovation systems were thus arranged in a very horizontal pattern (branch-related). Keener competition, saturated and increasingly demand-dominated markets, but also rapidly growing costs for chemicals development and testing however soon required a complete re-orientation. Innovations have to be promoted at the latest since the 1980s and 1990s and right from the start in a more customer-related way, the innovation systems have to be arranged vertically along the supply chain. The communication along the supply chain, required already to improve the ability to be innovative, is additionally forced by REACH with regard to risk communication. Improved communication along the supply chain will also improve the prerequisites for securing brand strategies and for customer-oriented and better adapted solutions (customer retention, quality competition). Greater demand for risk-related information and/or relevant product qualities could thus constitute self-supporting quality competition, in which health and environment-related qualities are adequately taken into account.

1.5   Prerequisites for success in hazardous substance substitution

After this demonstration of higher-level correlations, some significant SubChem results are now to be presented together with the related recommendations for action in the form of highlights.

1.5.1   Substitution of hazardous substances is an innovation process like many others

The substitution of hazardous substances in products and processes by lower-risk alternatives is generally an innovation process just like others too. The substitution of hazardous substances shares most features of any innovation, such as uncertainty about its success, overcoming inertia, favourable or unfavourable time frames (dependence on investment cycles) or dependence on the commitment of individual persons.

1.5.2   Chemicals innovations concern not only new substances, but also new preparations and applications

The inventory of chemicals available on the European market comprises around 100,000 substances, of which about 30,000 are marketed in quantities > 1t/a per annum. The quantity of new substances registered every year is around 0.3% of the inventory of substances.

---

[14] This would then be similar to a typical "technology push" innovation.

- This low ratio is partly due to the relatively high data requirements for registering new substances (with small tonnages) and also the relatively early point of registration in the innovation cycle (before technical trials with customers).
- At the same time, however, development of a new application or new formulation of preparations on the basis of the available inventory of substances are important examples of innovation with a high added value in the chemical industry. In this instance many SMEs are active in developing new recipes or in adapting existing ones to specific customer requirements. This means that the ability to be innovative in the domain of chemicals cannot be illustrated solely by the number of newly registered substances. Or to say it in another way: simplifying notification of New Substances is not the sole and maybe not even the essential factor for the ability to be innovative.

1.5.3   Public and civilian society – combined with intensive competition – are powerful driving forces in the innovation system

Various types of markets, in which companies operate (supply-dominated versus demand-dominated; Fordist[15] mass market versus differentiated quality market) have a considerable influence on companies' opportunities and restrictions for action.

- The costs for additional product qualities in the environmental and health area can thus hardly be passed on in the market for mass products with industrial/commercial customers (construction materials, chemical base substances) if need be with latent surplus production, so long as these are individual precursors.
- In many cases, however, cost arguments (direct costs per unit) are considerably overestimated as far as their significance for the substitution of hazardous substances is concerned. It is rather the indirect costs (which are decisive for all forms of innovation), the expenditure on additional training programmes, the need to change operating processes or the amount of investment costs, which impede substitution.
- The proximity of the actors in the supply chain (manufacturers, formulators, dealers, end users) to the end consumer rather has the effect of encouraging substitution. Conversely, in areas that are remote from the consumer, one of the essential driving forces for elimination of hazardous substances is frequently lacking, i.e. the public and/or consumers' organisations. A critical public combined with state regulation (which frequently also reacts to public scandals) has so far created the most significant impulses for change and the speediest substitution processes in competition-intensive areas.

---

[15] The terms Fordist or Fordism are derived from the mass production and mass consumption market introduced by Henry Ford in the early 1900s.

- Contaminant scandals or product deficiencies in the broader sense that have been the subject of public debate do produce learning effects, which extend far beyond the product and company concerned. This means that journalistic interest in the subject of chemicals and comparative testing facilities also are important driving forces for innovation. At the same time, it is however also necessary to ensure that the effective instrument of "public opinion" is handled in a responsible way.

1.5.4   Competition is the most important driving force for innovation – the demand for quality may create a direction for it

Competition is the most effective driving force for innovation. In competition-intensive, quality-differentiated, saturated and demand-dominated markets the most important impulses for hazardous substance substitution should originate from the customers, both from the end customers and from industrial or commercial chemicals users:

- More intensive communication along the supply chain and better knowledge of users' interests are a recipe for success, not only for hazardous substance substitution, but also generally for the market position and for innovations in the chemical industry. A vital forming option in the chemical products market for professional users is to consider dialogue along the supply chains as a chance for more customer-oriented innovation. The scheme for the new European chemicals policy reinforces these impulses of the globalised market by creating a standard mechanism for communication about substance properties and application conditions in the supply chains. Potential for economic success can thus also be exploited.
- Particularly in industries structured in small enterprises with limited users' know-how and a wide range of chemicals applications, the field service of preparations manufacturers or the trade have a central role in conveying product information and in advising on applications. In particular the chemicals business and the formulators could also expand their business areas to include information and consultation services (with the associated customer retention).
- Hazardous substances will (without introducing bans) only disappear from the market (and thus from products and articles) to the extent that, firstly, users demand aspects of environmental, health and consumer protection as qualities and, secondly, secure themselves against scandals involving hazardous substances by adopting precautionary strategies. The incentives to demand environmental and health-related information and to select the most compatible product based on that are thus intensified. In this respect "green and safe products" are also more attractive to the commercial user for various reasons. These reasons include the following aspects i.a.:
  - changes in orientation and wishes of end consumers (more "chemical product safety");

- sensitive reactions of share prices to headlines and warnings, and related further demands of shareholders with regard to risk protection and management;
- desire as a company to achieve good benchmarking (especially with regard to ecological and health objectives);
- increasingly corporate-culture-based, proactive styles in environmental and health issues, frequently also in connection with the uncertainty surrounding developing legal dynamics;
- ongoing public discussion on the subject of chemicals and the influence of associations (environmental, consumer, trade unions);
- efficiency gains and savings in classic environmental and occupational health and safety costs, e.g. disposal costs, efficient use of materials, avoidance of complex measures for employee protection.

1.5.5   Harmonised rules are needed for risk communication in the market

A more marked acceptance of responsibility by the market actors – such as within the future framework of REACH – requires the development of new forms of communication and co-operation along the supply chains. If substance-related information (by substance manufacturers) and user-related information (by preparations manufacturers and/or users) are not collected in the supply chains, the market actors remain dependent on assessments by state bodies.

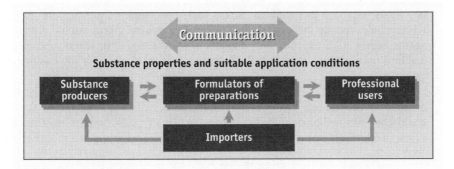

Figure 2. Communication along supply chains

Responsible communication in the chains throughout the European market in turn requires a common European standard to determine, evaluate, document and communicate substance risks. This also includes communication concerning transparent rules of procedure with the aim of giving legitimacy to the substance assessments and the decisions taken on the basis of them. This is especially important where defining and limiting risks has to be "curtailed" for pragmatic reasons in view of limits to knowledge. The market actors will have to assume themselves the responsibility for this "curtailment" of substance safety assessments within the framework of the REACH system.

1.5.6 Understanding complex innovation systems improves the ability to be innovative

For joint action in innovation systems and especially also for outlining the framework conditions by the political actors, it is expedient to understand the innovation systems. The model developed in the scope of the project serves to create a systematic link between the framework conditions, the influential factors and the correlations between the participants. The various contributory parties can better assess their own options for influence, existing resistance to new ideas, possible coalitions or the significance of market trends, and can approach change processes in a more purposeful way as a result of this.

The constellations of actors in the supply chain may be structured comparatively simply and in a linear way, but may also be interlinked and complex. The more complex the network of actors and the more extensive the particular innovation (level of innovation) proves to be, the more difficult the innovation process is and the more marked either the external impulses or the intrinsic motivations of the participants in the chain must be. With regard to the complexity of the innovation system, two basic system types have been deduced into which each of the 13 case studies contained in the SubChem project can be assigned. As ideal types the following differentiations are made:

- comparatively straightforward systems, in which the cause-effect relations can be clearly assigned to individual actors and their range of instruments. This type includes mostly the case studies of cement, mineral fibres, mould release agents, where it is frequently only a matter of substituting a particular substance and thus maintaining the technical effectiveness of the product for the commercial user.
- highly complex and dynamic systems, in which only the interaction of a large number of participants can produce innovations, which no individual actor could have planned or predicted (emergence). This type includes mostly those cases, in which

  - perceptions of future consumer wishes play an important role,
  - no clear system leader in the chain can be identified,
  - the supply chain is globally interlinked and
  - the chemical products are integrated in a complex machinery environment,

  such as is the case with textile auxiliary agents.

If this approach to type classification is successfully differentiated further, an instrument for structuring innovation processes can be developed on the basis of it.

1.5.7   Risk reduction too has to contend with conflicts of objectives

Hazardous substance substitutions are often pursued not far enough or in a wrong direction:

- In many cases simply substituting one substance for another is not enough. Generally technical innovations must accompany organisational and institutional innovations. The change processes must be organised between the employees or various departments within a company. Moreover, co-ordination between various companies is required or a favourable time frame has to be adapted. The "new" substance may not hinder combination with other components.
- Without simultaneous changes in the product or process design, substituting substances is the best solution only in the rarest of cases. This becomes apparent in the cases where the substitute possibly had less known, but by no means less dangerous properties, e.g. as the chemical structure had hardly been changed with regard to unchallenged requirements for substance properties. This is the case for example for some substitutes for PCBs and for man-made mineral fibres as a substitute for asbestos.
- Risk avoidance often faces conflicts in objectives. In this way substance innovations for improved fire and explosion protection in the second half of the last century also made a major contribution to many environmental and health risks discussed today: CFCs, PCB, chlorinated solvents, brominated flame retardants and also cutting oils in emulsion form must be mentioned at this point. This means that chemical/technical innovation originally aimed to achieve more operational safety. Additional improvements related to environmental or health aspects were not taken into account at the time of the innovation.
- Substitution is generally linked with considerable assessment problems. It is important to avoid shifting the risks and also to keep track of both the substance and energy flows as well as resource consumption under sustainability aspects. Examples of such shifts include accident risks in the substitution of chemicals for high-pressure facade cleaning, fire risk in the substitution of cooling lubricant emulsions by oils, discharges to water and the increase in contact allergies when converting solvent-based or mineral-oil-based systems to water-based systems.

1.5.8   Dealing with the "lack of knowledge" is a key to innovation – extended risk management is required

As long as no agreement can be achieved in public about the limits of 'burdens of evidence' and/or about the realistic extent or principal limitations of scientific knowledge of the effects of chemicals, there will be repeated attempts to prevent any type of innovation using the irrefutable claim as to uncertainty and "lack of knowledge". Innovation and risk are, however, inseparably interlinked. Not only economic and technical, but also social, health and ecological risks are always linked with innovations. What is missing are simple methods of assessment concerning the extent of uncertainties and the extent of possible consequences, i.e. where applicable, indications of an extreme 'level of risk' (for detecting possibly 'intolerable' or non-accountable risks), and/or risk management which takes into account the lack of knowledge in an appropriate way[16].

- Knowledge of the possible (eco)toxicological effects of chemical substances in complex chemical-biological processes is inevitably limited. Neither the available economic and ecological resources nor the objective of animal protection that is rapidly gaining importance at present allow all possible (eco)toxicological properties and all possible conditions to be tested using standard laboratory experiments. This means that the required bases for perception and decision-making not only for risk management, but generally for any kind of practical action in the development and selection of chemical substances for processes and products must be defined and organised so that actors are not forced to wait for extensive "toxicological certainties".
- As risk management cannot be based solely on knowledge of toxicological effects, strategies for rationally dealing with uncertainties are required. More certainty in orientation, for example, could be produced by a national chemicals strategy, such as that which exists in Sweden or in The Netherlands. In Germany the corresponding quantitative objectives, time targets and guiding principles have so far been missing.
- Substitution of hazardous substances is an element of risk management. In this context it is among the more demanding options of risk reduction. It is important to develop as rational strategies as possible for dealing with the lack of knowledge (cf. Sections 1.5.9 and 1.5.10).

1.5.9 Guiding principles may provide orientation but may also be misleading

The direction of the change process and the orientation of the contributory parties play a key role in avoiding substance-related risks by substituting hazardous substances. When developing alternatives, it should be ensured firstly that shifts in risks do not take place (including other types of impacts). Secondly, there will always be uncertainty with regard to assessment related to the limited possibilities for predicting effects on ecosystems and effects on the human organism.

- Guiding principles, such as the "recycling management system" or "natural raw materials" for example, had in the past a considerable effect on the innovation processes and the public debate in the field of chemicals and/or substance economy. It must be presumed that this will remain the case in the future too,

---

[16] A review of those procedures, which were developed in companies to assess and deal with economic risks (risk management) reveals ways of applying them to the processes for dealing with social, health and ecological risks.

although the influence of guiding principles is set to increase rather than to decline.

- Guiding principles related to substances and techniques may provide orientation, but only if they contain a reference to the risk. Thus under risk aspects generally accepted, positive guiding principles like "natural substances" or "water-based products" can thus result in certain risks being underestimated (e.g. the human toxicity of natural substances). Conversely, the guiding principle of "chemistry of short range" (chemicals with low environmental persistence and transport distance) or "intrinsically safe products" is apparently sufficient to provide orientation for developing and designing substances and products and still not to curtail the risk assessment inadmissibly.
- Not least of all it is a matter as to what extent the new product or process quality can be communicated to potential customers and is approved of or even causes "enthusiasm". Guiding principles can play an important role in this case. In addition to the instruments of regulation and economic control, they are among the most interesting approaches from the group of 'informational' control instruments (3rd generation).

1.5.10 Simple rules for risk management are needed

A vital objective of future European chemicals policy is to avoid uncontrolled handling of hazardous substances. Assessment of each individual application on the basis of the individual substances involved (as provided for by the current regulatory system) is not an apt strategy, as the number of assessment and management cases that it produces is much too high. Quantitative risk analyses are only taken into account in the case of substances with clearly definable effect thresholds and controllable application conditions. For 90% of the market actors the particular product or application system must possess "intrinsically safe" properties, as most companies have neither closed systems nor the required skills to deal with hazardous substances.

So, simple management rules are needed to ensure the basic levels of protection. Examples of such rules could be as follows:

- No application of persistent, mobile and/or bioaccumulable chemicals in open systems. Risk reduction by substitution or system shut-down is required.
- No application of CMR substances[17] or other highly dangerous substances by commercial[18] users outside industrial installations, as experience demonstrates that the required protective measures are not observed in too many instances. Risk reduction by substitution is needed. In the case of closed industrial installations, especially for chemical synthesis, minimisation of technical emissions (without substitution) may be the better strategy.

---

[17] Carcinogenic, mutagenic and reprotoxic substances
[18] For preparations for use in private homes a general ban on use already applies.

Substitution strategies must also take into account the users' situation in an appropriate way and must adapt to simple assessment instruments and to risk management employing simple rules. An elaborated, differentiated and self-responsible risk management, which has long since been self-evident for the major companies in the chemical industry, is an excessive challenge for many commercial companies, which process or use chemicals. Both in the areas of occupational health and safety and also in substance-related environmental and consumer protection comparatively demanding risk management plans are still required; simple decision and information instruments (or those graded in accordance with the problem dimensions) are still too underdeveloped and are hardly widespread. Interesting examples of existing "simple" instruments are the British COSHH Essentials[19], the control banding scheme[20], the Swedish Observation List[21], product codes such as the GISCODE for products in the construction industry[22] or the product classifications used in the TEGEWA system[23].

1.5.11 The entire range of opportunities to exert influence by the state has to be exploited

All in all the state framework for determining, assessing and communicating chemical-specific risks in the supply chains of the European market has an important significance. Localising responsibilities with the market actors at the individual stages of the supply chain, harmonisation of demands and standardisation of instruments are the essential challenges.

However, exclusively regulation-driven substitution of hazardous substances does not work. Approaches based on general regulations, e.g. the substitution principle contained in the Ordinance on Hazardous Substances, are unable to take action up to the level of small and medium-sized enterprises or crafts users due to structure-related deficits in enforcement. This means that additional driving forces are needed. However, the legally established obligation of industrial or commer-

---

[19] UK Health and Safety Executive: COSSH ESSENTIALS – Easy Steps to Control Chemicals; HSG 193; May 1999

[20] Den Umgang mit Gefahrstoffen in Klein- und Mittelbetrieben sicherer gestalten, Workshop presented plans for chemicals management at the workplace, October 1999, www.baua.de/news/archiv/pm_99/pm83_99.htm

[21] Swedish National Chemicals Inspectorate: Observation List – Examples of Substances Requiring Particular Attention; second revised edition 1998

[22] Code system for chemicals used in the construction industry, based on the hazardous substance information system of the statutory accident insurance in the construction industry (GISBAU); cf. Berufsgenossenschaften der Bauwirtschaft: Gefahrstoffe beim Bauen, Renovieren und Reinigen; Frankfurt, 2001

[23] Verband der Textilhilfsmittel-, Lederhilfsmittel-, Gerbstoff- und Waschrohstoff-Industrie (TEGEWA): voluntary commitment to classify textile auxiliary agents in accordance with their waste water relevance; November 1997

cial chemicals users to use lower-risk alternatives, if these are available, does appear to play an important role in individual cases.

Strategies of successful state intervention could also be as follows:

- Announcement effects or focused campaigns of enforcement agencies: if users of hazardous substances have to expect new regulations or the authorities initiate enforcement initiatives, there is a greater market chance for innovative products and services. In doing so, it is frequently more a question of creating a clearly visible impulse than comprehensive enforcement.
- Qualification of state bodies and other institutions with sovereign duties in such a way that they can perform consultative duties, especially with regard to SMEs.
- Establishment of mechanisms for cost internalisation (liability matters and insurance obligations): employers' liability insurance carriers can indeed affect risk management. Coverage against liability claims and linking insurance protection to minimum standards of risk management (production and product) are thus important driving forces for risk-reducing innovations with regard to consumer protection. On the other hand, mechanisms for internalisation in the area of chemicals-related environmental costs, which are caused by a continuous discharge of hazardous substances from various sources, hardly exist. Examples are the additional costs in providing drinking water, in the disposal of sewage sludge or in the refurbishment of buildings contaminated by hazardous substances.
- Development and application of a standard for "good assessment practice": The quality of risk assessments and risk management information could become an element of competition in the global market. To this end an auditable and internationally recognised standard is required.
- State institutions may support pilot and reference companies within the scope of innovation promotion. They may promote vertical but also horizontal communication (e.g. branch dialogues) and especially provide basic preliminary services for small and medium-sized enterprises, such as e.g. free offers of information about *best practices*, branch benchmarks or also initiate qualification programmes.
- Innovation is also driven by state-backed R&D programmes. A greater alignment of research promotion in accordance with guiding principles such as "intrinsic safety of products", "short-range chemistry" or "sustainable chemistry" may give direction to innovation efforts in industry.

2 Initial situation and analysis of deficits

2.1 Substitution of hazardous substances – introduction and definitions

Hazards and risks originating in chemicals are not a new phenomenon; they have appeared throughout human history right from the start. However, the economic/technical significance of 'chemistry' in modern industrialised societies is new, as is in this context the evolution of hazardous substances in relation to their quality (hazardous nature and diversity) and their quantity (volumes processed).

Some keywords concerning the production and consumption of chemicals[24]

- Since 1930 the quantity of globally produced organic chemicals has grown 400-fold.
- In the EINECS inventory, 101,600 different chemicals are registered.
- The EU Member States account for 50% of global chemicals trade, with 30% of exports going to non-EU countries.
- The EU foreign trade surplus for chemical products is around € 100,000 million p.a. The value of imports from Asia and Eastern Europe is € 16,000 million and growing.
- The number of European chemicals manufacturers is around 23,000 companies, with 95% of them employing fewer than 250 members of staff.
- The number of industrial chemicals users in the EU is around 500,000 companies.
- The number of consumers on the EU market will in future be around 460 million

---

Besides the dangers originating from hazardous substances and the conflicting interests of the involved parties, the discussion surrounding chemicals policy is also characterised by problems of comprehension, e.g. between corporate practitioners and proponents of regulation theories and/or between experts and the public. We thus present some definitions at the start of this chapter, at least for the sake of improved terminological clarity:

---

[24] EIA (2003)

- In terms of chemicals legislation **substances** are elements or compounds in the natural state or obtained by any production process, including any additive necessary to preserve the stability of the products and any impurity from the process used. A "substance" (as determined by its CAS or EINECS number) may be a desired combination of products from a chemical reaction or a distillation fraction (in refinery products) and/or may contain undesired impurities. This means that the same "substance" may indeed exhibit different properties.
- **Hazardous substances** are chemical substances with properties that may have certain harmful effects on humans and on the environment upon contact or exposure. The risk arises from the degree of hazard presented by the substance (including any of its metabolites) as well as the nature and intensity of contact (duration, dosage, absorption route, frequency). Hazardous substances in a legal sense are chemicals which must be classified as hazardous due to certain predefined criteria and available information or which have been classified as hazardous by official bodies.
- **Preparations** are mixtures or solutions comprising two or more substances.
- **Articles** differ from preparations or substances by the fact that they have a specific form or surface. The function of the product is determined by this form (e.g. sheets of floor covering) and less so by the composition of the substance (e.g. polymers and additives).
- **Substitution** generally means that one substance or process is substituted for another, for whatever reason (availability, costs, technical requirements). Substitution of a hazardous substance or product signifies its replacement by less a hazardous substance, product or process. In this context the scope ranges from simple substitution (i.e. exchanging substances) to risk management as a whole (i.e. prevention of hazardous substances, reduction or prevention of exposure, etc.).

The objective of the "SubChem" project was to analyse both the current situation and the framework conditions of hazardous substance substitution and also to draw up options for future strategies for action. The time frame being considered and the required review of individual case studies was generally limited to approximately twenty years.

Many aspects of substitution, its conditions and effects already described in existing subject literature were confirmed in the course of work on the case studies, but this was mostly only in certain cases and without any systematic framework for interpretation. This includes the enforcement deficit for application-related regulations in small and medium-sized enterprises, the obstructive behaviour of powerful associations or the trend towards more public attention with regard to substances in consumer products – in comparison with commercially used preparations or substances.

However, it was not until the last decade that some framework conditions changed significantly. This led to other problems and other attempts at finding solutions than those that were common in the 1970s and 1980s.

What framework conditions have changed significantly?

The regulation spectrum has been extended, especially in the area of environmental protection. In occupational health and safety the substitution of hazardous substances at this time became *de jure* a priority obligation in the area of risk management, *de facto* it remained ineffectual. Since 1981 there has also been an obligation to notify new substances with the authorities using a standardised set of original data, before placing them on the market for the first time. However, substances tested before being marketed even today account for less than 5% (in terms of number of substances) or 0.1% (in volume terms) of the entire market. Around 15 years later approximately 12,000 existing substances (with a production/import volume > 10 t/a) together with the data available in industry were logged in an inventory within the scope of the European existing substances regulation[25] (constitution of IUCLID database). So far less than 1% of this inventory of substances has been evaluated systematically and comprehensibly for the authorities and the public.

Within the scope of the new European chemicals policy (REACH system) a new attempt is now being made to reduce the information and assessment deficit with regard to the hazardous properties and risks originating from existing chemicals.

The public interest in safe and healthy products or at least products not hampered by scandals is growing. In the area of environmental protection the first 'chemicals pioneer period', which was perceived almost completely positively by the public, has long since past and a strict level of regulation – at least for averting hazards and also as an initial step to taking precautions against risks – has generally been established.

At the same time, economic constraints, growing unemployment and highly intensified international competition may indeed act as barriers for economically not entirely assessable substitution projects. Retaining on established procedures, economically successful, but in part ecologically problematic existing substances is favoured by this. This is the one side. On the other hand, companies in intensive competition depend much more on the basis of and with a good reputation. In public all fluctuations and image changes are registered much more quickly. The 'vulnerability' of companies has grown infinitely. In view of fragmented markets and shorter product cycles, market success is closely interlinked with the ability to make innovation (innovation push).

2.2   The case of asbestos – introduction to the subject

Some central and typical aspects of hazardous substance substitution are to be illustrated using a prominent and as yet unfinished example – the substitution of asbestos. The absolutely never-ending story of the use, regulation and substitution of asbestos is one of the best known 'cases' of substitution of all time.

---

[25] Council Regulation (EEC) 793/93 or the Existing Substances Regulation (**ESR**)

Have possible warning signals been ignored?

Asbestos is a natural fibre, which from a purely technical aspect is ideal for a great deal of applications, particularly insulation, friction pads such as brakes and couplings, filtration and fire protection. Its physical structure, the length of the fibres in combination with their stability and longevity, make asbestos so effective – and at the same time so dangerous. This indicates that certain properties such as above-average persistence to biological decomposition with corresponding exposure options could – or should – give rise to a fundamental scepticism in evaluation. This is also an important approach for substitution. The composition of some of the fibres used as substitutes is as bio soluble as possible – which is meant to be both a simple and an ingenious solution.

Scientific cognition – necessary but 'inadequate' by far

In the 1920s there was already clear medical evidence that work-related contact with asbestos is detrimental to health[26]. Despite this, global asbestos consumption continued to grow from approximately 300,000 tons in 1925 to more than 5 million tons in 1975. It was not until then that the most dangerous asbestos applications – e.g. spraying of asbestos cement – were banned and protective measures were stepped up.

> *'Looking back in the light of present knowledge, it is impossible not to feel that opportunities for discovery and prevention of asbestos disease were badly missed.'*
>
> Thomas Legge, ex Chief Medical Inspector of Factories, in Industrial maladies, 1934.
>
> quoted in: *Asbestos: from 'magic' to malevolent mineral*; European Environment Agency: Late lessons from early warnings: the precautionary principle 1896-2000, Copenhagen 2001

The 'indestructibility' of asbestos is in this respect a scientific stroke of luck. The fibres can still be detected even years later; the combination of the cause 'asbestos' with the effect 'mesothelioma' was thus comparatively simple. Normally the correlations between occupational diseases and the use of hazardous substances are much less evident.

Sensitive public – high pressure on a small point

A relatively fast and extensive prohibition regulation, which was suddenly introduced, was greatly promoted in Germany in the 1980s by the fact that in addition to employees other groups – considered publicly to be more sensitive – spoke out. Although union initiatives, and in particular initiatives by shipyard workers, had highlighted the hazards of asbestos on several occasions since the early 1970s, the real break-through did not come until studies established that also children in

---

[26] Late lessons from early warnings, EEA (2001)

schools and nurseries, teaching staff and the public at large were at risk due to asbestos contained in asbestos cement.

Employees are apparently relatively unsuccessful when mobilising the public for their health interests. The phenomenon existing in the labour world of considering health risks to be a challenge to working life and thus almost to be an 'enhancement of one's own activity' has been described in many instances (cf. box). The view of 'what does not kill us will harden us' is one of the more negative framework conditions for substitution.

The British industrial doctor Lucy Deane wrote the following in 1898 concerning the behaviour of workers with regard to asbestos:

"Even when the evil reaches such grave proportions as to be capable of easy and tragic proof... there is always a certain proportion of 'old workers' – the survivors of their mates – who are found in every unhealthy industry and who... appear to thrive on their unhealthy calling. In less obvious unhealthy conditions the only convincing proof of actual injury, viz. reliable comparative statistics of mortality, or of health standards, is practically unattainable in the case of any given factory, or at any rate with the time and opportunity at present at our disposal."

in: European Environment Agency: Late lessons from early warnings: the precautionary principle 1896-2000, Copenhagen 2001

A decisive factor for the success of substitution was even in this case the ability of those concerned to reach and sensitise the entire public, i.e. society at large. This frequently happens concerning the level of concern of especially sensitive or also prominent groups. The - from a toxicological aspect - 'ludicrously' small concentration of TBT contained in the jerseys of a German Football League team, which attracted more public attention than reports about the effects on joggers, textile workers or textile retailers could ever have done, once again demonstrated such effects in an extreme way.

Globally standardised procedure – no chance?

Another problem is the extent of regulative measures. On the one hand, the use of asbestos is now almost completely banned in the European Union and in many other industrialised and economically thriving countries. On the other hand, more than 2 million tons of asbestos are extracted and consumed globally every year. A uniform 'global' hazardous substances policy evidently does not exist yet.

This may be due to the economic situation in many countries, which extract or consume asbestos (catching up on industrialisation). In Germany also there were stages of extensive economic primacy, e.g. during reconstruction in the 1950s and 1960s. During this period of rapid industrial growth asbestos was not an issue. Between 1940 and 1973 not a single preventive regulation relating to asbestos was implemented. However, in 'modern industrialised nations' also the direct economic interests may become prevalent, resulting in entrenchment on highly dubious positions – this is the case for asbestos e.g. in Canada.

When price dictates the decision to buy

Asbestos was a relatively economical product for a mass market of commercial customers. In other markets better quality also achieves a better market price. This quality may also mean for customers being free of hazardous components (textiles, cosmetics, children's toys).

Stable mass markets, as in the case of asbestos or in some of the case studies (cf. Chapter 4.2) such as cement or DEHP, are generally less innovative. This is especially the case if the product is indeed 'already existing', but is technically superior and at the same time economical as the result of established mass production.

Avoiding innovation stress

Also in the case of asbestos a phenomenon regarding those involved became apparent that is also always difficult to comprehend, even for "SubChem": a low ability to be innovative and a high level of entrenchment. The interest in innovation of the actors involved is in most instances overestimated from the outside – i.e. within the scope of a scientific examination. The ability to become entrenched is however underestimated, with inertia doubtless being a significant barrier. Voluntary innovation efforts are considered to be unnecessary additional stress in established routines.

This observation is however not restricted to supposedly conservative factory owners, who at best fear liability problems or financial risks. It applies to workers and technicians alike, to purchasing department and management, to associations and scientific branch institutes as well as to sub-contractors of machinery, components and auxiliary equipment, in which the chemicals are used. This conservative stance applies not only with regard to environment and health-related innovation, but also to innovation in general and it is amplified (and also repeatedly upset) by the stress of globalisation and social uncertainty.

Aftercare instead of precaution

The phenomenon that the use of hazardous substances was not restricted for a long time despite early clear indications as to the risks involved exists not only in the case of asbestos, but also in the case of other substances such as lead, many solvents, CFCs, mercury etc. Only in the period of marked industrial growth (1950s, 1960s, 1970s) did the use of many of these hazardous substances reach a level that led to tangible consequences (e.g. as in the case of CFCs), large-scale contaminations (lead, PCBs) or occasional disasters (in the case of mercury and dioxin). In addition, typical delays are observable repeatedly with the result that one or several decades can pass between the cause of damage, the corresponding perception and the taking of sustainable action. Aspects of such delays include:

- latency until damage becomes visible;
- internal scientific discussion of cause/effect correlations;
- slow leaching into public attention and political sphere;

- development of processes for systematic measuring or prediction of new types of effects;
- assessment of results in the conflict of interests;
- development of alternatives;
- regulative measures with corresponding transition periods.

Search process involving risks

By the end of the 1980s more than 3000 substitutes for the various asbestos applications had already been developed or had generally been adapted from other applications. For most of these substitute materials and products the level of risk for workers and the exposed population is considerably lower. In this respect substitution was 'worthwhile'.

However, at a later stage some of these substitutes have also in turn been identified as being hazardous, e.g. carcinogenic, with the result that new substitutes had to be searched for. At the latest since this example the fear is always present that one evil is being replaced with another in the form of the substitute, when one hazardous substance is substituted by another that possibly entails new risks. There is clearly no complete certainty and security of direction in substitution. As in other social areas too, action needs to be taken despite gaps in knowledge. Neither insisting on 'complete cognition' nor exaggerating 'fundamental deficits in knowledge' is appropriate for this situation.

And in the case of asbestos, as in other cases, other 'proponents of precautions for risks' are frequently opposed to further restrictions: associations representing the fire brigades have for a long time opposed a ban.

Externalisations of costs – the state is the re-insurer

The overall economic costs of the health consequences of asbestos were assigned to society at large. The German legal system does not make a clear assignment to those responsible for such costs. The costs are high: at present around 1000 workers die in Germany every year as the result of exposure to asbestos; the association of commercial and industrial employers' liability insurance carriers expects 20,000 deaths by 2020 (cf. figure 3). The overall costs including accident benefits will then rise to more than € 10,000 million[27].

---

[27] Dr. Joachim Breuer, CEO of the association of commercial and industrial statutory accident insurance, stated in a press release on 17.12.2002: "We do not expect deaths to reach their peak until between 2005 and 2015". Breuer estimates that the overall expenditure of the iability insurance carriers could probably rise to well in excess of € 10,000 million. By 2020 up to 20,000 deaths in Germany could be due to asbestos.

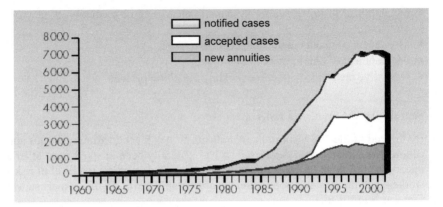

Figure 3. Industrial diseases caused by asbestos[27]

The few asbestos-processing plants do not pay for these costs, as they would presumably only have been able to do so with a very good third-party liability insurance, which for its part would have caused the product price to increase sharply. These costs are paid by the community of user and non-user plants in a particular industry via the employers' liability insurance in the construction industry and the employers' liability insurances in the metal industry).

In many other cases of work-related exposure to hazardous substances, however, the cause/effect correlations do not permit such unequivocal proof. The number of acknowledged industrial diseases caused by hazardous substances should thus be much lower than the number of persons actually affected. Thus, the risks of compensation are accordingly low for the employers as the contributors to the employers' liability insurance schemes.

2.3   New general framework conditions

How do the general framework conditions for the substitution of hazardous chemicals differ from the conditions under which the case of asbestos was dealt with? This analysis was so important for the SubChem project because the relevance of the historic case analyses for future substitution processes (cf. Chapter 3) had to be assessed.

2.3.1   Perception of environmental and health issues

Three objectives can be identified as higher-level objectives of regulation – both for chemicals and also with regard to other environmental and health risks (biohazards, physical risks and risks of radiation) – according to the environmental guidelines of the German government in 1986[28]:

a. Averting dangers: Averting imminent environmental or health risks (accidents, known critical exposure or emissions with direct harmful consequences in most cases);

b. Risk precaution: Averting measures for risk prevention and reduction taken in the initial stages. Risk precaution thus means also taking into account possible damage, "which cannot be excluded only because certain causal relations can neither be confirmed nor negated in accordance with the current level of knowledge and so no danger exists at this stage, but rather only a suspicion of danger or potential for concern."

c. Future precaution: Measures for the protection and development of the natural basis for life. *"Future precautions are best served by developing production processes and products, which do not produce substance emissions that are a burden to the environment and are detrimental to health or at least avoid these as far as possible."*[29]

These general objectives should form an extensive consensus. There is likewise a general consensus that environmental, occupational, health and consumer policies over the past thirty years have achieved considerable material success in reducing the use of known hazardous substances, also with regard to their loads in soil, air, water and products and not least of all with regard to their safe handling at the workplace.

Nevertheless, when analysing the initial situation of the substitution of hazardous substances in 2003, recourse to documentation from the 1980s, the "German government's guidelines for environmental precaution by avoiding and gradually reducing hazardous substances" are in fact a stopgap measure and are rather unusual. This indicates a lack in updating and specifying of a social guiding principle or a "national strategy" – or even better a "European strategy" – for chemicals policy.

In the course of its 340 pages the German government's sustainability strategy for Germany[30] does not mention even once the term "chemicals", neither in the glossary nor in the chapter headings. This national sustainability strategy does not contain any future-oriented ideas in relation to chemicals policy.

The intensity of the discussion concerning the new European chemicals policy, involving both industry and NGOs, is in contrast an indication that chemicals policy continues to have considerable potential for conflict. A consensus between industrial policy on the one hand and health and environmental policy on the other is nowhere to be seen.

---

[28] Guidelines for environmental precaution (1986)

[29] Ahlers, J. et al.: (2001)

[30] German government: Perspektiven für Deutschland: Unsere Strategie für eine nachhaltige Entwicklung (Perspectives for Germany: Our strategy for sustainable development), (2002)

Unlike the discussion on environmental policy in the 1980s, support for a strict continuation of environmental and health policy appears rather to have waned. After some scandalous 'classic' hazards were reduced, actions for precautionary risk reduction again appear to be weighted more against the requirements of industrial and economic policy. This is also reflected in the fact that the latest initiatives by the EU Commission in the area of environmental and chemicals policy were accompanied by an intensive discussion of the "correct" method for assessing the consequences of legislation.

Despite acknowledged successes, some outstanding issues still exist regarding the actual consequences of exposure to chemicals for humans and the environment, which have become more pressing:

Synergy effects: What does it mean that humans and the environment – compared with the 1960s – are today less exposed to known hazardous substances, but that the diversity of chemicals contained in consumer products (frequently with an inadequate data status) has increased?

Systemic effects: Are certain more systemic, multifactor health risks (such as e.g. allergies, hormone changes, certain types of cancer) favoured? The rise in the incidence of allergies in the European population was e.g. one of the main justifications of the EU Commission for the need to improve chemicals policy.

2.3.2   Scientific range of instruments

Progress in analysis as well as in toxicology and ecotoxicology has doubtless also contributed to intensifying the public debate concerning the risks relating to hazardous substances. The analytical methods and the range of instruments used in the sciences involved in substance and risk assessment have become much more differentiated and also more comprehensive. From the metrology aspect alone, much more precise substantiation methods have been developed. Further end points have been added from the domain of toxicology (e.g. hormone system, wildlife); the entire life cycle of a substance can be described on the basis of life cycle analyses.

2.3.3   From banning to range of control instruments

Bans and strict application regulations were still viable options following the asbestos debate in the 1970s and 1980s. Today the range of state actions (or governance) is much more broadly defined. The three classic forms of regulatory policies-

a) authorising (burden of proof with companies),
b) regulating the application and
c) banning (burden of proof with authorities) – are complemented today by diverse actions, for which state, semi-state and private institutions frequently co-operate with each other in different ways:

Co-operative legislation: In view of the increasing complexity of economic processes and technology, legislators have increasingly to rely on economic information in order to draw up the regulative framework conditions in a practicable way. On the other hand, the economy needs state enforced rules in order to facilitate fair competition, while at the same time protecting both health and the environment.

Voluntary agreements and voluntary-commitments: In Germany there are currently more than 30 voluntary agreements and voluntary commitments by the chemical industry trade associations the objective of which is to substitute (i.e. phase-out) or reduce certain substances[31]. Of this figure, approximately 20 agreements relate to such hazardous substances, which are used internally and to this extent are relevant for occupational health issues. Such voluntary-commitments include, for example, the commitment by the coatings industry not to use certain heavy metal compounds in coatings and paints as well as its obligation to reduce solvents (1984)[32], CFCs in detergents and cleaning agents (1987)[33] or the raw materials exclusion list for printing inks, primers, coatings and overprint varnishes (1995)[34].

Branch agreements: In addition some limited branch agreements do exist, for example for reducing solvents used in offset printing (Bundesverband Druck – Federal Association of German Printing Industry, IG Medien – media workers' union, machinery manufacturers, sub-contractors) and also relating to low-chromate cement (employers' liability insurance in the construction industry, employers' associations, construction industry workers' union, construction materials trade etc.)

Provision of assessment aids, guidelines and databases: State or public bodies have developed new forms of supporting the application of laws. These include primarily electronic database systems or Internet-based applications (GISBAU, GISCHEM, GESTIS, MALEG system)[35]. Other countries are also active in this field; the best known example is e-COSHH in Britain. e-COSHH is an internet-based instrument to enable in particular small and medium-sized enterprises to handle chemicals as prescribed by regulations; this instrument also contains pre-defined solution options for standard operating situations[36]. On behalf of the German Umweltbundesamt (Federal Environment Agency) Ökopol developed a "Guidance for the use of environmentally sound substances"[37]. These guidelines

---

[31] Federal Institute for Occupational Health and safety/Saxony-Anhalt: Voluntary commitments by the associations of the chemical industry, status 15.07.01

[32] Lißner, L: (2000)

[33] http://www.umweltbundesamt.de/uba-info-daten/daten/wasch/anforder.htm#CKW

[34] BDI: www.bdi-online.de

[35] Cf. www.gisbau.de, www.gischem.de, http://www.hvbg.de/d/bia/fac/stoffdb/index.html, www.maleg.de

[36] Russell, R. M. et al. (1998)

[37] Guidance for the use of environmentally sound substances, for producers and professional users of chemical products relevant to the aquatic environment (2003), cf. http://www.umweltbundesamt.de/umweltvertraegliche-stoffe-e/guidehome.htm

offer assistance to small and medium-sized enterprises in interpreting and evaluating available substance information with the result that various technical options can be compared more effectively with regard to their environmental compatibility. This type of instrument also includes "grey or black lists of substances", which already offer orientation before regulative intervention of the economy: Sweden and Denmark thus published lists containing substances that are due to be subject to stringent regulations in the foreseeable future ('List of undesirable substances' and 'Observation list' respectively)[38]. Similar lists are also maintained in large multinational companies or in certain industrial associations (cf. automotive industry[39]).

Steering charges to reduce undesirable behaviour

In the field of hazardous substance reduction these instruments, which are known from the area of environmental protection, for example in the form of waste water charges, have so far not been employed in Germany. In Europe there are some examples, such as the VOC[40] charges in Switzerland for each litre of solvent used, the penalty charge on perchloroethylene and pesticides in Norway, or the Danish tax on phthalates in soft plastic. Some regulations from the area of waste management – for example, separate storage and compulsory return of chlorinated solvents – did in fact have a similar effect indirectly by making the use of certain substances much more expensive by increasing the costs for waste disposal.

Labelling (and relevant test prerequisites)

The labelling of negative or positive characteristics of substances and products aims to use market mechanisms to force hazardous substances out of the market and to create incentives for more environmentally compatible and health-compatible products. Caution labels (such as 'dead fish beside dead tree' pictogram or the skull symbol) and positive signs ('environment angel') can give rise to substitution, provided that they are backed up by appropriately consistent criteria. Private labels, such as Ökotex 100, or marking systems, such as the ARS system used by German manufacturers of textile auxiliary agents, or the GISCODE used by the employers' liability insurance carriers in the construction industry apply to this kind of summarised positive/negative hazard related information. Finally, the future certification of construction products in accordance with the Directive for Construction Products[41] is also included in this group of instruments.

---

[38] List of undesirable substances 2000, Danish Environmental Protection Agency, DEPA, http://www.mst.dk/homepage/

[39] E.g. International List of Reportable Substances (ILRS) of the IMDS (International Material Data System: www.mdsystem.org

[40] VOC = Volatile Organic Compounds

[41] Council Directive 89/106/EEC of 21 December 1988 on the approximation of laws, regulations and administrative provisions of the Member States relating to construction products (OJ No L 40/12 of 1989-02-11)

2.3.4   International integration

Global chemicals trading and the global organisation of supply chains also require harmonisation of the relevant rules for evaluation, communication, documentation and limitations on use. Under such conditions, isolated national solutions are scarcely feasible, are in most cases legally controversial and are also not practicable. On the other hand, the international contractual system extending beyond the EU is tedious and requires national or regional antecedents. Companies operating globally can also assume pioneering functions, such as the automotive industry with its International Material Data System (IMDS). For all actors, both economic actors and national authorities, international integration beyond the supply chains represents a huge challenge.

2.3.5   Substitution has priority – unfortunately not in practice

With regard to dealing with hazardous substances, a great deal has been achieved in the regulative area since the Ordinance on Hazardous Substances came into force in 1986. A clear substitution principle was devised in this ordinance:

§ 16, Para. 2: The employer must check whether substances, preparations or products with a lower health risk than those he intends to introduce are available. If it is reasonable for him to use such substances, preparations or products and if substitution is necessary to protect the life and health of employees, only they may be used.

In the meantime this substitution principle has also been applied in EU legislation. "Council Directive 98/24/EC of 7th April 1998 on the protection of the health and safety of workers from the risks related to chemical agents at work" compulsorily provides for substitution as priority action:

Article 6

(1) The employer shall ensure that the risk from a hazardous chemical agent to the health and safety of workers at work is eliminated or reduced to a minimum.

(2) In applying paragraph 1, substitution shall by preference be undertaken, whereby the employer shall avoid the use of a hazardous chemical agent by replacing it with a chemical agent or process which, under its condition of use, is not hazardous or less hazardous to workers' health and safety, as the case may be.

In actual fact neither such general regulations (substitution principle), nor prescribed procedures (risk assessment) nor also complex detail regulations (Technical rules for hazardous substances, TRGS) have caused a major push towards substitution. On the one hand, the general regulations are considered to be ineffective, yet on the other hand the detailed regulations are especially difficult to manage and are not so relevant to practical conditions.

2.3.6   Enforcement deficits

Reduction of exposure to hazardous substances as the result of protective measures is still a primary concern. Technical, organisational or personal protective
measures are prevalent in occupational health and safety. Technical emissions
control as end-of-pipe measure is traditionally the principal approach in the area of
environmental protection. The systematic substitution of hazardous substances being used as a control measure in company risk management is, however, still of
secondary importance. Most problems in dealing with hazardous substances arise
in any case in the many small and medium-sized enterprises that use chemicals
with inadequate know-how and protective measures, in many cases openly and
manually[42]. The same applies to the required management structures for producing
high-quality safety data sheets in compliance with the legislation and for using this
information source. Here too the implementation quota in companies is insufficient[43]. All in all it can be assumed that around 70% of commercial users of hazardous substances do not (or cannot) observe the statutory requirements of employee protection[44].

This situation does not apply only to Germany. The Health and Safety Executive quotes specific figures for the UK, which give an idea of application of occupational health and safety legislation there[45]. In approximately 1.3 million British
companies chemicals are handled. When questioned, only 16% of these companies
were able to state the applicable law for handling chemicals or the limit values for
these substances at the workplace. With regard to the need for protective measures, these 16% also rely almost entirely on the details contained in the safety data
sheets.

The (non) observance of complex regulations when dealing with chemicals is a
known and widespread problem for enterprises in all EU Member States. With regard to the status of health and safety in the context of chemical risks at the workplace the monitoring study of the European Agency for Health and safety at Work
states in a laconic way: "There is a need for monitoring compliance with legislation"[46].

By way of reaction to the evident deficit in implementing the differentiated
chemicals/hazardous substance legislation in Europe, the authorities responsible in
all EU Member States set up the CLEEN network (Chemical Legislation En-

---

[42] Voullaire and Kliemt; BAuA publications series (1995). Hazardous substances in small
and medium-sized enterprises – Proceedings of the information symposium in Dortmund
on 7th/ 8th May 1996; BAuA publications series; Tb 75, Dortmund/Berlin 1997

[43] Wriedt, H. et al, 1999

[44] Study by Ministry of Social Affairs of the State of Hesse: Status report and future of
ASCA – Evaluation of results of 670 corporate studies on the organisation of occupational health and safety and implementation of occupational health and safety regulations; Wiesbaden, 2000

[45] Topping, M., HSE Health Directorate, UK: 2002

[46] European Agency for Health and safety at Work: Monitoring the State of Occupational
Health and safety in the European Union – Pilot Study, p. 41

forcement Network). This network implements projects to identify the shortcomings in application of legislation and to develop measures for improvement. The applicable reports are available to the public[47].

2.3.7   Many substitution incentives originate from waste and waste water legislation

In the past twenty years many legal provisions have been created to regulate substance flows (recycling management systems, waste management, electrical and automotive recycling). In many instances these laws explicitly contain threshold values for certain substances or even ban certain substances. They are thus very effective on the use of these substances in production processes. This is also true for threshold values of chlorinated compounds in industrial waste, the ban on certain heavy metals in the automotive industry and substance-related requirements for waste water from the textile industry (Annexe 38 of Waste Water Ordinance[48]).

2.3.8   Markets and brands

As a result of globalisation and market saturation, competition has intensified considerably and will continue to do so. Production sites where wages are significantly lower, including sometimes also lower environmental and social standards, impose pressure upon prices and costs. The European manufacturers cannot react to this pressure merely with cost-reducing and rationalisation strategies. Gaining new market shares and securing existing market shares by quality differentiation must also be practiced. Competition purely on the basis of cost would hardly be successful. In the final analysis, it is important to gain the upper hand not only on the basis of price competition, but also on the basis of quality competition with the other industrialised nations. With comparable technical qualities and comparable prices, the sales-relevant qualities can indeed also include health-related and environment-related aspects of product and process planning.

It is important to offer customers a direction in the variety of competing products in order to retain customers' loyalty to the company. Brand strategies have thus gained huge importance for sales success in consumer-near areas. Brand names are, however, more sensitive to factors threatening their image than no-name products are. Brand names are thus also "sensitive" to environmental and health 'reproaches'. This was not yet the case for asbestos producers in the 1970s, as their names were unknown outside specialised groups. However, most consum-

---

[47] NONS (1996), EUREX (1999), EuroCad (2001), SENSE (1998), SENSE (1999), ECLIPS (2004)

[48] Ordinance on requirements for discharging waste water in waterways – Waste Water Ordinance, 15th October 2002

ers are familiar with the names of international brand textile producers, chemicals companies or automotive groups.

The economic framework conditions for hazardous substance substitution have thus also changed considerably in the course of the past twenty years. Many commodities are produced today in globally organised supply chains; the origin, quality and application conditions of process chemicals and product components is complex. Developing appropriate management systems for product quality and product safety requires considerable organisational innovation by the companies involved.

2.3.9   Innovation

The outsourcing of BAYER Chemicals (today LANCESS) from the BAYER Group in early 2004 may be interpreted as an expression of innovation expectations and thus yield expectations by the company management: with a wide range of organic specialities competitive margins on the capital markets can no longer be obtained. Instead, the innovation efforts focus on *"Life Science"*, *"Crop Science"* or *"Material Science"*.

The domain of the German chemical industry, on the basis of major combined production sites and fully integrated synthesis chains to supply chemicals to the industry is being superseded by increasing specialisation and concentration on certain product and demand segments. In addition, e.g. organic dyes and certain pharmaceutical preliminary products are now being produced in Asia considerably cheaper yet in adequate qualities with the result that imports to the EU are also increasing.

One indication of innovative activities of the chemical industry is the development of patents. The proportion of organic chemicals in the entire patent history of chemicals was around 10% in 1920 and reached its maximum of around 40% by the end of the 1970s. Since then the proportion of organic-synthetic chemicals in innovation history in Germany is declining again. The relative significance of other areas, such as e.g. pharmaceuticals, biochemistry and materials research is growing accordingly[49].

Since 1981 around 3000 new substances have been registered in Europe, yet have rarely reached significant market volumes. Every year around 300 new substances are notified. According to the new substances statistics of the European Chemicals Bureau so far only 0.6% of substances have a market volume in excess of 1000 t/a. and a further 2.9% have reached a market volume of between 100 and 1000 t/a.[50]. Most new substances are specialities, mainly in the areas of semi-finished goods for synthesis, process regulators, dyes and photographic chemicals[51].

---

[49] Grupp et al. (2002)

[50] http://ecb.jrc.it/new-chemicals/

[51] Substances to be evaluated in accordance with the biocide directive as per a comparable process since 2003 are not dealt with here.

Most innovations appear to concentrate on more efficient manufacturing processes, new formulations (preparation innovation) and the transfer of preparations
to other areas of application (application innovation).

- Innovations in process technology did play a key role in the past. In this way
  many substances can be produced with a higher yield and a lower energy consumption today.
- The same applies for preparations, i.e. the combination of various substances in
  a product with a certain technical performance. In this field a high level of innovation dynamics still appears to be prevalent. To what extent these are 'mere'
  optimisations by maintaining and modifying the recipes or are really new products, with an innovative technical performance, is difficult to determine.

An essential ingredient in the innovation history in the area of chemicals is,
however, doubtless the development of recipes for preparations and their specific
adaptation to the requirements of fast-moving manufacturing processes, for example in the automotive, electronics, printing and paper industries. This occurs
mainly on the basis of the existing inventory of primary materials.

Concerning the ability to be innovative, however, possibly innovation-
inhibiting effects of state chemicals regulation were and still are discussed critically on repeated occasions. For example, Fleischer (2000) and his team compare
the registration of new substances in the USA with the rate of new substance registrations in Europe over the past 20 years. On the basis of this indicator they
come to the view that the registration system in the USA produced more new substances by 1999 (i.e. is more supportive of innovations) than the European registration system is[52]. Nordbeck and Faust[53], however, rightly point out that the differences in the registration rates over the past 10 years have declined considerably,
which may be related to the fact that the European register of existing substances
(100,000 substances) in 1980 was almost twice as large as the US register of existing substances (62,000 substances). The consequence is as follows: The "inventory of substances" that did not need to be registered in Europe was larger than in
the USA, i.e. the ratio of new substances grew more slowly and did not stabilise
until the second half of the 1990s at a level of 300 per annum. It must still be
stated that the production or marketing intake (> 10 t/a) in the case of new substances in the USA is about 1.75 times higher than the registration of new substances (> 1 t/a) in the EU[54]. For these figures the question remains unresolved as
to where substance development itself takes place, whether the US market acts as
a test market for substances developed in Europe and whether substances produced (but not marketed) in Europe are first registered for marketing in the USA.

---

[52] Fleischer et al. 2000;
[53] Nordbeck and Faust, 2002
[54] Fleischer, 2002

2.3.10 Trade

The chemicals trade has gained in significance like trade as a whole. Chemicals exports have tripled since 1980 (in absolute turnover figures), while imports have quadrupled[55]. Many synthesis processes, which twenty years ago were still a domain of German and European industry (e.g. organic pigments), can today be performed in India or in Asia.

The chemicals trade has gained in significance due to the increasing complexity of products in the inter-trade sector also. Companies in the chemicals trade, which frequently also have a department manufacturing their own formulations, assume a key role between substance manufacturers and SME users. However, the chemicals trade cannot always be identified as such, e.g. the DIY branch which has grown substantially since the 1980s sells major volumes of consumer-near chemicals (mineral wool, paints, wood preservatives).

2.4 Characterisation of the current EU regulatory system

The criteria and procedural methods for characterising substances with regard to their hazardous characteristics and risk characterisation are harmonised in Europe. There also exists a common European system for the classification and labelling of hazardous substances and for safety data sheets. Officially harmonised classifications also exist for around 3000 substances. The individual elements of the current system as well as its weaknesses are to be explained in the following section.

Quantitative dimension

The inventory of substances on the EU market is just over 100,000 substances listed in the EINECS register (European Inventory of Existing Chemical Substances[56]) and ELINCS register (European List of Notified New Substances[57]).

According to estimations by the European Commission, of the 100,000 EINECS substances around 30,000 are manufactured or imported and marketed in the EU in annual volumes in excess of one metric ton. Of this figure, around 2500 substances have a production or import volume exceeding 1000 metric tons per annum. For these substances considerably more ecotoxicological and toxicological data are available than averaged out at all substances[58].

---

[55] Verband der Chemischen Industrie (VCI): Chemiewirtschaft in Zahlen 2003, p. 80

[56] Inventory on the EU market 1981, according to information supplied by the chemical industry.

[57] Marketed as new products since 1981.

[58] Institute for Health and Consumer Protection – European Chemicals Bureau: Public Availability of Data on EU High Production Volume Chemicals, 1999. These substances are documented in the IUCLID database (International Uniform Chemical Information Database).

The number of new substances notified in the past twenty years is around 3000; every year about 300 substances are now notified. The fraction of substances achieving a production volume in excess of 1 metric ton is around 70%.

Hazardous substances include not only synthetic chemicals, but also some natural materials such as mineral (dust or fibres) and a range of metals. It is very difficult to estimate the production volume and distribution of these materials.

2.4.1   Instruments for substance evaluation and risk information

The four basic instruments for substance evaluation and hazard information can be summarised as follows:

- The classification and labelling of hazardous substances and preparations as per a harmonised system has been done in the EU since 1967. Over the past 35 years this system has been continuously refined; in particular the health effects of chronic exposure and the environment-related hazards have been gradually integrated. It must however be taken into account that chronic human-toxic effects can frequently only be researched by means of extensive animal testing, which has not been carried out for most substances to date. By international comparison the EU has a very extensive list of hazardous substances with harmonised classification containing around 3000 entries. By 2008 the EU intends to introduce the established system to the GHS (Globally Harmonised System) within the scope of the UN and adapt it accordingly.
- The harmonised EU safety data sheet is designed to describe systematically the hazardous properties of chemicals and to inform users about suitable and effective risk management measures. For some years this safety data sheet also has to state what (eco)toxicological standard effects have not been verified for a specific substance. The EU instrument has now been in existence for thirteen years, while the corresponding DIN predecessor had already been introduced in the 1980s. Nevertheless, many safety data sheets still indicate extensive quality shortcomings.
- The systematic evaluation of substance properties and predictable or actual exposure patterns over the entire life-time of a substance within the scope of risk assessment is as yet a relatively recent instrument, for which harmonised scientific rules were created in the EU for the first time in 1997 in the form of the Technical Guidance Documents (TGD). An essential element in this range of instruments is how to deal with shortcomings in knowledge. Wherever information is missing, standardised worst-case scenarios are conceived taking into account appropriate safety factors[59]. If under these worst-case assumptions a rele-

---

[59] Ibid.: The Technical Guidance Document Part I states (page 60): "To decide the reasonable worst case, all measured data sets and qualifying information gathered for a particular scenario should be considered. Since the aim is the identification of highly exposed workers, the aim is to try to approximate to the 90th percentile values. When the data set is small, values near the highest end of the concentration range should be used, to ensure that highly exposed workers are represented. In this case, critical evaluation of outliers,

vant risk results, the assessment can be refined by collecting data closer to reality or, alternatively, suitable measures for risk management have to be derived. In practice this set of rules has so far been applied mainly by official bodies, by the applicants for biocide and pesticide authorisations as well as by those registering new substances. The manufacturers of industrial "existing substances" (= EINECS substances) apply a similar set of rules at most when assessing substances that have been prioritised by the Member States for existing substance evaluation. So far no European *Guidance Document* exists for substance-related risk analysis at workplaces, as was stipulated by EU Chemical Agents Directive 98/24/EC.

• The obligation of each manufacturer or importer to provide a binding data set on the chemical-physical and (eco)toxicological properties of substances before marketing (fundamental or basic data set) is existing in the EU as an instrument since the early 1980s, however it has so far only been applied to new substances.

System of EU risk assessment

With its Technical Guidance[60] the European Union together with the Member States outlined prerequisites for assessing substances (or a "soft law", as it is known).

The outcome of the risk assessment carried out in accordance with this creates conclusions in four possible graduations for new substances (in the case of existing substances three graduations). The essential conclusions concern the need for additional information (for example, additional tests) or ascertain that action is needed to reduce the risk (cf. original text contained in the following figure 4)[61]:

---

to determine whether or not they should be excluded, becomes essential to avoid an unrealistic estimation."

[60] Institute for Health and Consumer Protection – European Chemicals Bureau: Technical Guidance Document on Risk Assessment, European Communities 2003

[61] Ibid., Part 1, p. 10

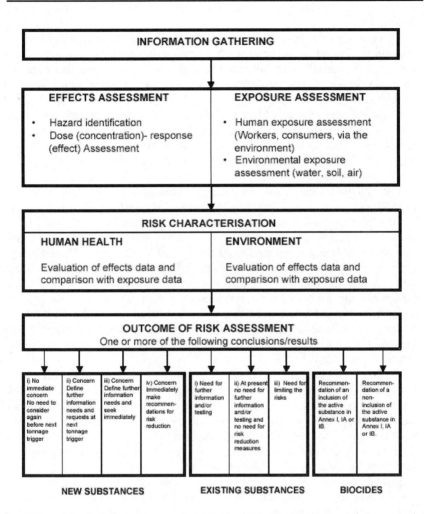

Figure 4. General principle of risk assessment for new substances, existing substances and biocides

Assessment of environmental and health risks is based on the dose-response assessment and also on the systematic estimation of the exposure levels along the various exposure pathways.

To determine the risks for individual substances in the environment, the concentration of a substance is determined in the individual environmental compartments – water, soil, air – (on the basis of measurements or calculations). This measured or calculated concentration (PEC = Predicted Environmental Concentration) is compared to the concentration at which no detrimental effects on ecosystems are expected (PNEC = Predicted No Effect Level Concentration). The PNEC is ascertained from laboratory tests on individual organisms and appropriate safety factors.

To determine the risks in the area of human health (workers, consumers, including the absorption of chemicals from the environment) the concentration of a substance is ascertained for so-called "end points" in the human body (organs or biological systems). The measured or calculated concentration (exposure level) is compared with the NOAEL (No Observed Adverse Effect Level). The NOAEL is the highest concentration at which no more effect is observed (usually in animal testing), i.e. there is no visible or measurable effect. If such a "level" cannot be detected, the LOAEL (Lowest Observed Adverse Effect Level) can be applied instead. This is the value at which the effect first becomes visible or measurable with an increasing dose.

For genotoxic and highly sensitising substances neither a NOAEL nor a LOAEL can be derived, as no threshold value for triggering the effect can be determined for these substances. In addition the effects, unlike caustic and irritant effects, may occur with a considerable time delay. This means that in the case of such substances any exposure should be avoided at all, if possible. Furthermore, carcinogens cause severe chronic health damage even at very small concentrations. Within the scope of the new European chemicals policy risk assessment procedures are suggested for substances with such properties; these procedures differ from the standard procedures, which are based on threshold values. Up to 5% of industrial chemicals found on the market today show these types of properties and should therefore not be used at all or, if used, only in highly controlled 'closed' applications.

Persistent and highly bioaccumulative substances are just as difficult a case, albeit in a slightly different way. Persistence and high mobility (volatility) combined with either fat- or water-solubility and bioaccumulability in any case indicate high environment-induced exposure. In this case too it is hardly possible to predict what emission loads can lead to health damage in the long term, for example in the case of mammals of the sea. Measures for precautionary risk management must start here, i.e. reducing or avoiding exposure.

In reality many of these very hazardous substances occur not only in highly controllable applications, but also in working and living environments, in foodstuffs and in ecosystems, in many cases far from the actual place and time of their planned purpose of use.

New substance assessment

The main objective of new substance assessment is to ascertain whether objections should be raised to marketing of a new chemical.

Before a new substance is first marketed, the 'placer on the market', i.e. manufacturer and importer, has to notify it and to present proof that not risk arises from the use of this substance. In Germany the Notification Unit of the Federal Institute for Occupational Health and safety (FIOSH) receives the relevant data.

The scope of data sets to be submitted for registration depends on the marketing volume of the substance. If new substances are to be marketed in a quantity > 1 t/a., test results have to be presented in accordance with a legal catalogue of basic data[62]. In the case of larger quantities additional tests must be presented, in the case of smaller quantities more simplified data sets are sufficient[63]. On the basis of the substance data sets those uses must be evaluated, which are intended by or known to the manufacturer/importer.

Existing substance assessment

The objective of assessing existing substances is to ascertain the risks involved in substances that have already been marketed and to reduce these, if necessary. Any restrictions on marketing and use that may be required represent a market intervention and therefore have to be substantiated well by the authorities.

In accordance with EEC ordinance no. 793/93 (existing substances ordinance) dated 1993, manufacturers and importers have to submit the available data for every substance listed in EINECS to the European Chemicals Bureau (ECB), if their production volume or import volume was in excess of 10 t/a in a predefined reference period. All in all, the industry has submitted data covering the basic data set of European chemicals legislation in only 14% of cases for high-volume existing substances (2500 substances > 1000 t/a)[64]. Details concerning exposure were even more patchy.

At the beginning of 1993 the European Commission in collaboration with the Member States likewise selected 140 priority substances for a Community assessment programme for existing substances in the EU. Processing these priority substances has so far progressed very slowly. In total approximately 45 concluded and complete risk assessments (March 2004) exist. 17 of these final risk assessments[65] led to the conclusion that risk management measures are required on Community level. This conclusion is backed up by a second process on EU level, within the scope of which the required action is set out in specific terms. This process is also slow. So far regulative action has been taken only for a few substances. The overall situation is described by Ahlers et al. as follows[66] "Unlike new substances, for which defined, effective data drawn up in accordance with internationally recognised guidelines are submitted, in most cases a very large, heterogeneous (and frequently not standardised) and often contradictory data set exists for the existing substances being assessed. This forms the basis for forming an overall picture of the hazards of the individual environment compartments. In many cases essential basic data are also missing."

However, it must also be pointed out that the existing substance assessment programme has in the meantime resulted in widespread harmonisation of methods

---

[62] Article 7(1) together with Annexe VII A of Directive 67/548/EEC

[63] Article 7(2) together with Annexe VIII Stage 1 or Stage 2 or Article 8 together with Annexe VII B, C or D of Directive 67/548/EEC

[64] Allanou, Hansen, van der Bilt, 1999

[65] http://ecb.jrc.it/existing-chemicals/: Existing Substances Regulation – Results

[66] Ahlers, J. et al.: 2001

between Member States. This is a good basis for making the system more efficient.

Summary of central deficits in the present EU system

The structural deficits of the present regulatory system can be summarised in five points:

- Existing substances and new substances are still being treated differently by applicable regulations even 25 years after introduction of the EU chemicals legislation.
- Many measures in the areas of employee protection and environmental protection are not initiated until a substance or a preparation is classified and/or characterised as being hazardous. If existing data are not adequate to assess a substance, this initially has no consequences. This means that the present system leads to completely inappropriate and initially inconsequential gaps in knowledge for existing substances. It rewards the substance manufacturer, who does not invest in collecting substance data, structurally.
- For more than 95% of substances on the market the responsibility for the risk assessment of substances is with the authorities and not with the substance manufacturers.
- The duty of substance consumers to carry out risk assessments for employee protection is generally not applied on the basis of the present range of instruments.
- The attempt to carry out risk assessment on a scientific basis has resulted in the systematic separation of risk assessment and practical risk management. An undesirable effect of this strict separation is the fact that the present practice of risk assessment does not produce the information that is required for formulation of risk management measures in a goal-oriented way.
- The existing substances ordinance only introduced obligations for substance manufacturers to provide information to the authorities, but not for substance consumers. This resulted in systematic gaps or practice-remote worst-case scenarios in exposure assessment.

Consequences of structural deficits for substitution

The comparatively small number of new substances that are marketed and the high import ratio of new substances from outside the EU zone (approximately 40%) indicate that the legislative unequal treatment of existing substances and new substances in the past two decades has had an effect of slowing down substance innovations[67]. Substances from the inventory of existing substances are mainly used for the substitution of hazardous substances, once this has in fact taken place.

Frequently hazardous substances cannot be compared systematically with the available alternatives due to the asymmetric data status. This slows down and prevents substitution processes, or results in incorrect suppositions concerning the

---

[67] Fleischer et al., 2000

"safety" of a particular alternative. Consequently entire series of misguided substitution efforts can be produced for the same product type.

It is notable that the differentiated possibilities for risk identification have comparatively little effect on the sale of existing substances on the market, while new substances cannot dominate a market, if they already give rise to comparatively limited toxicological misgivings when they are brought onto the market. One possible explanation for this phenomenon is certainly to be found in the economy. Developing new substances as an alternative to high-volume (and technically tried and tested) existing substances is only economically viable, if the existing substance is under marked and stable image pressure (as the result of assumed or actual risks) and pilot markets willing to pay exist, which pay the significantly higher price for the small-volume, new developed alternative.

3 Approach to research and procedure

Substituting hazardous substances by other substances is a self-evident strategy to reduce risks or already measurable damage caused by chemicals. The case of asbestos not only illustrates the fact that many more prerequisites have to be met for substitution actually to take place. The scientific partners of the SubChem team, comprising members from the Hamburg University of Applied Sciences (HAW), the University of Bremen (Department Production Engineering), the Ökopol-Institute in Hamburg and the Co-operation Centre in Hamburg[68], are aware of several examples of successful and also of extremely tough or even failed substitution processes as a result of their lengthy experience in the fields of environmental and employee protection.

On the basis of this experience, for the one part, and the results of innovation research driven by occupational health and safety and environmental protection to date, for the other part, the research tactic was developed to examine the driving forces and the restrictive factors as well as the appropriate framework conditions of substitution processes using case studies with the objective of deducing recommendations for action for present and future substitution cases.

The SubChem project was developed in the context of discovered deficits in regulating, assessing and dealing with chemicals in Germany and Europe. The project work coincided with review of the EU chemicals policy with the result that the initial scientific results of SubChem were also able to contribute to the political process surrounding the chemicals White Paper and the REACH ordinance.

3.1 Starting point in innovation research (driven by occupational health and safety and environmental protection)

Innovations driven by occupational health and safety and environmental protection possibly differ from other innovations in the objective and direction of innovation. For innovation processes, in which innovations driven by occupational health and safety and environmental protection are produced, certain general framework conditions and certain constellations of actors may be particularly favourable. Whether or to what extent the results of general innovation research dif-

---

[68] The Co-operation Centre is a division of the research department of the science and research authorities of the City of Hamburg.

fer fundamentally from the results of innovation research focused on environmental innovations, however, is and remains an unresolved question. In an initial approach it could be assumed that environmental innovations occur not solely or not primarily driven by competition and the market. So the important role of state measures in environment policy must be stated for the creation of environmental innovations also as a result of the BMBF-funded research project 'Innovative effects of environmental policy instruments' (FIU)[69]. These measures are not solely of a regulative nature, but rather also include instruments of the market economy (ecological tax reform), the promotion of research and development as well as supporting information and moderation measures. In this context the value of 'announcement effects' of state action was also singled out.

Klemmer et al. make the following conclusion: "Almost all studies reach the conclusion that environment innovations are the result of a more complex than one-dimensional sample of effects, which is formed by both intrinsic motivations and also by state incentives, characterised by numerous feedbacks and is largely dependent on the overall social environment: partly dictated by the design of the study, it is demonstrated using the example of selected sectors and/or groups of actors for the one part, and different environment policy problems for the other part, that there is no instrument to be favoured a priori or to be rejected generally and that only the interaction of individual motivations, political action and social environment leads to environmental innovations (multi-impulse hypothesis)" (Klemmer[71]).

They thus reach the following recommendations for action: "There is a broad consensus that it is not or not always the consistent application of individual instruments, but rather their choice and application within a certain sample of regulations that are the deciding factors. As a sample of regulations (referring to Jänicke 1997) the 'sum of all calculable rules, procedures and action contexts in a subject area of state control' is understood; this comprises the political-institutional context for action, the structure of instruments and the style of policy upheld in each case. According to these factors a successful, innovation-orientated environment policy is characterised by a style of policy, which

- already includes the various groups of actors on the level of objective forming and takes into account that changes in the constellations of actors require under certain circumstances other actions than those planned,
- chooses appropriate time scales and deals with the planned range of instruments in a flexible way,
- demonstrates a certain level of continuity and calculability" (ibid. p. 112[71]).

Precisely in the last formulations the reception of results of general innovation research becomes clear, which consequently also played an important role in designing the questions in the :riw-project. In many ways the 'Research initiative on sustainability and innovation – General framework for innovations for a sustainable economy' (:riw), in the context of which the SubChem project was also promoted, ties in with the findings of FIU and certain findings of general innovation

---

[69] Cf. Klemmer, P.; Lehr, U.; Löbbe, K 1999, Jänicke, M. (1997), p. 111

research. These are in particular findings about technological and economic path dependencies, i.e. the particular difficulty of changes in the technological path and 'basic innovations' as well as the 'depreciations' of fixed and variable capital that are always linked with innovations (sunk costs especially with regard to investments and qualifications). Recommendations for action orientated around windows of opportunity can be deduced from these findings[70]. Other approaches refer to especially favourable market conditions, niche markets and 'lead markets', which (sometimes also 'regulatively constituted') offer particularly favourable conditions for development for (environmental) innovations[71].

The research approach of SubChem also resorted to a widespread approach of general innovation research, i.e. 'thinking in innovation systems' (cf. chapter 5). The complex web of actors, constellations of actors, general framework conditions and influential factors in innovation processes can most appropriately be structured by a system-theory approach. The constellations of actors directly involved in the process act 'within the system' and the regulatory push and pull effects and market pull effects as well as the science-technology push are part of the general framework conditions. By way of addition to this more or less established 'crown' of general framework conditions, as the project progressed, there were also the 'public interest groups affecting this' as another particularly important influential factor.

With regard to the location of innovation systems, however, neither the regional nor the national level (national and regional innovation systems) could be applied in the SubChem project due to the subject of examination. A branch approach (sectoral innovation systems) was also not an appropriate way to approach the SubChem project. The innovation systems examined by SubChem are to be found over the entire supply chain from natural raw material supply and chemicals production to disposal and recycling (innovation systems going beyond supply chains and/or product life cycles).

3.2 Hazardous substance substitution as an innovation process

The SubChem project is concerned with preventing and reducing (eco)toxicological risks as a function of innovation within the scope of 'sustainable economic activity'. The processes of hazardous substance substitution are at the centre of interest[72]. They are examined as innovation processes, or to be more precise, as more or less 'targeted' innovation processes committed to 'environment' and 'health' values.

---

[70] Cf. in particular the SUSTIME project.

[71] Cf. in particular the LEAD-MARKET project

[72] "Hazardous substance" is used here in accordance with the criteria of German chemicals legislation irrespective of the fact whether a legal classification or manufacturer's classification exists.

As with each innovation process not only technical modifications but also new production processes and modified product features have to be taken into account. Also and especially economic and organisational changes, investments and new business relations as well as new assessment questions present considerable challenges for the actors in the innovation system.

As described in the initial situation of innovation research (Chapter 3.1), Sub-Chem used the system view and examined substitution as an innovation process in the innovation system, in which the actors are granted some leeway for action. The economic actors, who act within the supply chain, are directly involved in the process (cf. Figure 5). The political and economic actors outside the supply chain, who only influence the product flow indirectly, form the general framework conditions, which in turn characterise the actors' conditions for action within the supply chain (cf. Figure 1). Public interest groups and non-governmental organisations operate via the media and the public by assessing and highlighting these general framework conditions or the products in order to affect the actors' behaviour, e.g. consumer behaviour, in this way.

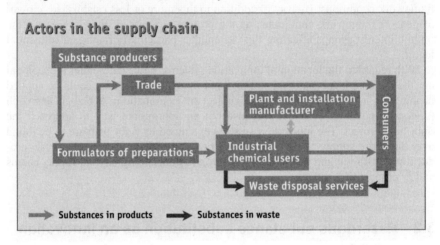

Figure 5. Actors in the supply chain

The behaviour of the individual actors in the chain is in turn affected by both internal and external structures and factors. For example employers, employees, purchasing departments for substance input and waste management for output as well as the various specialised divisions in companies using chemical products thus pursue entirely different interests and strategies. A central concern of the SubChem project was how the interaction of all actors inside and outside the supply chain can accelerate or hamper innovations.

3.3   Procedure used by SubChem

3.3.1   Research questions

The research work of SubChem was determined by three higher-ranking questions:

- How can economic, state and public interest groups interact successfully in innovation systems (*ability to be innovative*)?
- How can economic actors achieve greater certainty in orientation avoiding hazardous substances (direction of innovation)?
- How can substitution requirements be integrated in the internal (quality) management systems of companies and along the supply chain (*risk management*[73])?

In doing so, the first question essentially aims at gaining a better understanding of innovations related to hazardous substances in the supply chains in order to deduce indications as to how the actors within these innovation systems can optimise their efforts to minimise risk. The second and third questions are concerned with how the actors in the innovation system can achieve more safety with regard to the objective of preventing and reducing chemicals-related (eco)toxicological risks.

These two fundamental questions relating to the ability to be innovative and the direction of innovation basically structured the two parallel analytical work constituents in the project. On the basis of that, recommendations for action should then be formulated. With a better understanding of the ability to be innovative it is expected that both the actors in the system can optimse their actions, and that the actors outside the system can draw up proposals for improving the general framework conditions for innovations to reduce environmental and health risks originating from hazardous substances. Concerning the second question relating to the certainty in direction, the orientation itself changed in the course of the project. The third question was then raised only subsequently in this form. Originally improvements in risk assessment and risk evaluation were considered with regard to the direction of innovation, not lastly also with regard to an extensive sustainability evaluation, which must include much more than a 'classic' risk estimation and evaluation. In view of the fundamental limits to realisations of this project[74] – despite many possible improvements – the question shifted to risk management, the search for 'appropriate forms of dealing with lack of knowledge'. The questions relating to

---

[73] Risk management is understood in a broad sense in this context: i.e. risk prevention or risk avoidance, dealing with accepted risks (risk management in a narrower sense) or reduction of intolerable risks.

[74] Theoretical views concerning epistemiology and systems theory, ethical arguments (animal testing) as well as limited financial, time and human resources are important factors as reasons for these fundamental limits to realisation.

improvements in risk assessment and evaluation did not become obsolete as a result of this. The focus of the project work did, however, shift from the search for 'knowledge solutions' to the search for 'management solutions'.

3.3.2   Case studies, hypotheses, model: understanding of chemicals-related innovation systems

Innovation research is far from being able to work on the basis of a widely recognised structure of theories, the elements of which are just to be tested systematically and empirically. In many areas – and this also still applies to the questions that we pursued – it is firstly rather a question of generating and consolidating hypotheses about possible correlations and interactions which can be generalised. To this end the interplay between empirical case studies and theoretical modelling is an appropriate approach to research. The research work of SubChem was thus essentially founded on three elements: a) a preliminary understanding on the basis of own experience and former scientific findings, set out in a first set of hypotheses, b) a simple 'model' of an innovation system integrating the supply chain, c) 13 empirical case studies on both successful and also (so far) unsuccessful processes of hazardous substance substitution.

An essential prerequisite for processing this comparatively complex 'setting' was combining the networks of the three research partners within the scope of SubChem: university research (technology assessment, innovation research), corporate consulting (lengthy experience in the practical substitution of hazardous substances at the workplace), environment policy consulting and substance assessment (White Paper discussion, EU existing substance programme). The following aspects were of particular importance:

- utilisation of the range of risk management instruments from the domain of corporate occupational health and safety for environmental protection purposes;
- development of a common language and an integrated understanding of the problem between "public health advocates" and "environmental protection advocates";
- processing of company and intercompany everyday experience (which are frequently reported in an anecdotal way and interspersed with everyday hypotheses) using the analytical range of instruments applied in university research.

The empirical approach to the analysis of the case studies with particular focus on the users of hazardous substances was pursued in order to obtain a more extensive understanding of the practical possibilities and limits of precaution-oriented substitution of hazardous substances under the given general framework conditions, in various branches of industry and throughout very differing product lines. The whole process, however, did not only aim at making realisations, but also making recommendations for actions. Drawing up a 'system view' should, for example, allow actors in the supply chain to realise their own possibilities for action in innovation processes as well as their limits in order to obtain pointers about possibilities for influence and instruments, which have possibly been utilised insufficiently so far. The more the participants are able to adopt such a view of the 'entire innovation system', the earlier joint steps for improving the ability to be innovative should be attainable, i.e. steps towards a new quality of the entire system on the basis of individual contributions.

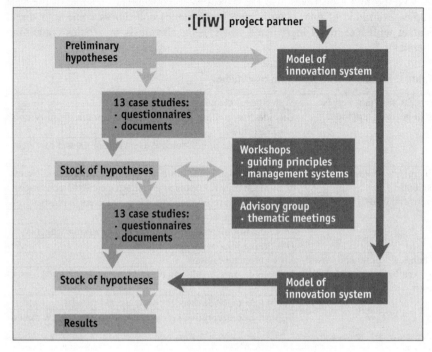

Figure 6. Working process and networking in the SubChem project

The case studies essentially fulfilled three functions in the SubChem project. Firstly, they served as a source of realisation; secondly as the background and illustration of hypotheses and type characterisation/modelling of innovation systems; and thirdly, for legibility to incorporate our results in "narratives about substances". The study proceeded on the basis of document analyses and collaboration with several practice partners in each instance. The cases were selected so that they covered as wide a spectrum of substitution conditions and product types as possible (cf. Table 1). In this way the cases covered the following:

- several focal subjects (subjects relating to environmental, employee or consumer protection) were covered;
- various competitive conditions (regional/global competition, supply or demand-dominated markets) were examined;
- various market constellations (long/short, regional/global supply chains) were considered;

- various users of chemical products (SME or large-scale industry, craftspeople, service-providers, DIY users) were consulted and
- both technical and organisational innovations were taken into account.

An important aspect was also the proximity to consumers. In some of the case studies examined (Table 1, with grey background) consumers come into direct contact with hazardous ingredients when using chemicals or articles containing chemicals.

Table 1. Summary of 13 SubChem case studies

Process auxiliary agents in industrial equipment	Water-based cleaning of metal surfaces
	Biocide-free cooling lubricants and minimum quantity cooling lubrication
	Bio-degradable mould release agents from natural raw materials
Chemical products/ components: industrial usage	Biosoluble, man-made mineral fibres in automotive silencers
	Man-made mineral fibres in automotive catalytic converters
	Alternatives to reproduction-toxic plasticisers in plastics
	Low-solvent automotive coatings
	More environmentally compatible textile auxiliary agents
	UV-drying inks in packaging printing
Chemical products: uses by craftspeople and DIY users	Low-chromate cement
	Biosoluble, man-made mineral fibres for insulation in structures
	Methylene chloride-free stripping agents
	Solvent-free decorative paints made from natural raw materials

Even at the start of the project a first, preliminary set of realisations (hypotheses) which can be generalised about possible success and failure factors in hazardous substance substitution on the basis of literature and the previous experience of the project group was developed. Using the empirical findings from the case studies, these hypotheses were examined, intensified and supplemented. Individual hypotheses or groups of hypotheses were also discussed and treated during the six-monthly meetings of the SubChem advisory board. We could take advantage of the varied experience of the members of the advisory board[75] concerning specific issues, which was also an effective contribution to quality assurance and result transfer. The interaction between case studies, hypotheses, model and advisory board meetings is illustrated in Figure 6.

In the original project plan, after processing all 13 case samples, a more detailed examination of two to three case samples and/or case groups was planned in

---

[75] The SubChem advisory board comprised representatives from associations (VCI, TEGEWA, DECHEMA, consumer advice centres, IGBCE), official bodies (FIOSH, Federal Environment Agency, Bremen health office), universities (University of Oldenburg, University of Augsburg, ETH Zurich) and also enterprises (Henkel KG, Volkswagen AG).

order to obtain more detailed findings about suitable management strategies and instruments for integrating hazardous substance substitution and prevention in the everyday actions of companies. As the project work progressed, it did however become evident that the essential issues concerning the direction of innovation could not be resolved solely on company level or by examining individual parts of supply chains. The same applies for ways of reinforcing the demander's position (cf. also model in Figure 7: Demand pull). Aspects and instruments beyond single case studies were subsequently treated more closely in the advisory board meetings In particular the perspectives of designing substances, processes and products oriented by guiding principles-and a wide concept of quality management spanning over the supply chain were discussed in two experts' workshops in autumn 2002 and 2003 (cf. Chapters 6.3, 6.4 and 6.5).

Together with the other chemicals-related projects in the :[riw] program (INNOCHEM and COIN) an abstract model of the innovation system was drawn up according to Hemmelskamp 2000[76]. To obtain a more generalisable understanding of chemicals-related innovation systems, the results of the case samples and the hypothesis development were also interlinked and abstracted in such a way that two basic types of innovation systems were able to be identified and illustrated towards the end of the project.

The supply chain and its actors lie at the heart of the model (Figure 7), which is affected by various "driving forces" (supporting or restricting factors). The system's ability to be innovative can thus be expressed as a constellation of actors with their possibilities for influence under certain framework conditions and in some cases a "basic atmosphere" of the system[77].

---

[76] Based on a model by Hemmelskamp et al., 2000

[77] This basic atmosphere of the system is frequently referred to as 'innovation climate'.

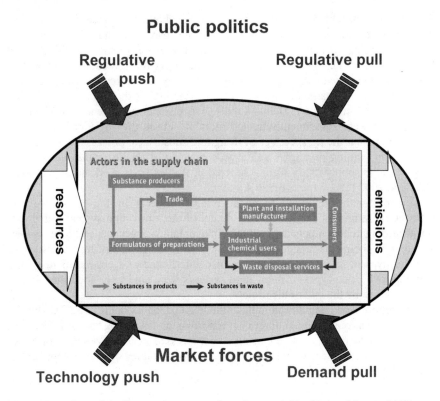

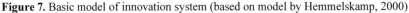

Figure 7. Basic model of innovation system (based on model by Hemmelskamp, 2000)

The most important driving forces include (in keeping with the results of the FIU project group[78]) the regulative push (left-hand red arrow) especially due to chemicals legislation (regulation related to introduction into market), the regulative pull by statutory environmental and health protection (plant and workplace-related and/or regulation related to application of products, right-hand red arrow), as well as the technological push by new substance/technical opportunities and market demand pull (blue arrows).

Types of innovation were used (substance innovation, product innovation, application innovation) and in some instances criteria for determining the level of innovation (chemical/technical and organisational/institutional, systemic) as a type feature for innovation systems.

Even if a preliminary understanding of innovation in the field of hazardous substances or chemicals tends to think initially of substance innovations, i.e. truly 'new' substances, as the project work progressed, the importance of developing new substances in the chemicals market (substance innovations) became clearly relative. When analysing the case studies, the choice of which had essentially been made from their relevance for the subject of substitution, it became evident how

---

[78] Cf. Klemmer, P./Lehr, U./Löbbe, K. (1999)

limited the role of actual "substance innovation" is in the case samples selected by us. In actual fact, the case studies are mainly new applications for (existing) substances already available on the market. None of the case studies was primarily concerned with the development of new substances. Some of our original hypotheses did, however, actually refer to the differing treatment of existing and new substances in legislation. In two of the case studies the aspect of substance innovations could also be examined more closely by expanding the products (or functions) examined. This was possible for the case studies "Alternatives to reproduction-toxic plasticisers in plastics" by including toys and medical devices and in the case of "Textile auxiliary agents" by including textile dyes.

3.3.3   Guiding principles and management solutions: orientation for actors in the innovation system

Pragmatic and speedy orientation in choosing between various substance/technical alternatives is at least equally important to the participants as improving their own possibilities for taking action. In too many instances the supposedly 'safer substitute' has later proved to be just as problematic or even more problematic.

Assessment of the direction of innovation has to contend with a whole series of fundamental problems, for which possible solutions however already are visible:

- Remaining lack of knowledge and remaining uncertainties even after applying scientific methods. One reaction to this is to out the precautionary principle into operation.
- The huge demand for resources and the expenditure of time, money, manpower (also not forgetting the suffering of laboratory animals) that is involved in the assessment procedures. In practice it is not the most sophisticated assessment procedures that are required, but rather pragmatic and practicable methods, which can produce a preliminary result in a 'rough and ready' way.
- In practice emotional driving forces frequently play an important role, as do positive or negative prejudices with regard to expectable spectrums of effects of certain substance groups. Scientific 'clarification' as a counteraction remains limited in its effects. It is much more a question of developing forms of comprehension and (as far as possible) rational ways of dealing with these emotional driving forces.
- Assessment of 'finished' chemicals takes place too late. (Subsequent) substance assessment has to be supplemented by guiding principle-oriented development of (intrinsically safe) chemicals and/or guiding principle-oriented forms of their intrinsically safe use.

For assessment of the substitution processes documented in the case studies with regard to the intended risk reduction (certainty of direction) various assessment approaches and strategies can be applied:

- Methodically devised and established methods (and criteria) of substance and process assessment and evaluation like risk analysis, toxicological and ecotoxicological analyses, life-cycle-analysis and cost-benefit analysis[79]. But also much simpler approaches, which in practice play an important role for assessing formulations and substance properties (e.g. the use of negative and/or positive lists).
- Criteria and speedy procedures for focussing the assessment on the risks that are probably relevant (preliminary processes, screening)[80].
- Assessment using substance-related guiding principles within the scope of substance development and substance applications.
- Creation of procedures for assessments, risk discussion and risk management

  - knowledge-based, scientific risk management,
  - precaution-oriented management against the background of lack of knowledge (not yet knowing and unknowability),
  - discussion-based risk management (inclusion of social valuations).

3.3.4   Elements of intervening research? – Realisations and influencing against the background of reorganisation of European chemicals policy

Work in the :[riw] program coincided with a far-reaching reform of European chemicals policy. This opened up extensive opportunities for not only an analytical, but also consultative (interventional) accompaniment of this extensive 'institutional innovation'. For SubChem this was primarily a matter of developing some fundamental correlations and founded (qualitative) arguments regarding possible ecological, health and economic opportunities and risks arising from the EU Commission's proposals for a new chemicals policy. Contributions for operationalisation of the precautionary principle and the economic effects of the REACH process to be expected were particularly in demand.

The:[riw] research program also offered a co-operation network, which had already been utilisable for this process and will continue to be utilisable beyond the scope of the project. It was already apparent at the :[riw] status seminar held in Heidelberg in May 2002 that the chemicals-related projects SubChem, INNOCHEM and COIN and the questions that they raised can make important contributions to the ongoing debate concerning the re-organisation of European chemicals policy. This is especially the case for the non-classic regulatory elements of policies contained in chemicals regulation. If we understand the development, manufacture, distribution and application of chemicals as processes within the scope of innovation systems, our attention is directed at the significant importance given to co-operation, information, communication and learning proc-

---

[79] Gleich, Arnim von (1999)
[80] Cf. Risk Commission 2003

esses along the supply chains, for the one part, and the actions of public interest groups and the public, for the other part.

As the political window of opportunity for institutional innovation in the system of chemicals regulation had opened parallel to the project work, linking the research project with the ongoing political process was of particular interest for those involved in the SubChem project. In doing so, the aim was to make the knowledge available in the :[riw] program accessible by participation in workshops, enquiries and specialised debates for the parties participating in the regulation process[81]. Due to the involvement of project members in various forums and events in the course of the REACH process and in the *Risk Commission* (10/2000-12/2003)[80] appointed by the German Environment and Health Ministries, there was an extensive interlinking of actors as well as an intensive exchange between researchers and politicians. On the other hand, the thematic workshops within the scope of SubChem were placed in a conceptual context with the re-organisation of European chemicals legislation. The question extending beyond REACH or continuing on from REACH as to how the economic actors will be able to fulfil the responsibility they are assigned or how they can adequately fulfil the role they are assigned in risk management was of particular interest. This of course again raises the question as to how the state institutions can best support them. In the final analysis this came down to the question as to how appropriate 'learning systems' (for single companies and also inter-company) can be organised (proceduralisation), as an alternative to fixed predefined rules of behaviour for the actors that are worked out in detail (cf. especially the workshop reports "Guiding principles" and "Risk management").

---

[81] The following must be emphasised in particular: involvement in the accompanying group of the Environment Ministry and participation in enquiries, conferences and specialised debates on the economic effects of the EU Commission's proposal for reform of European chemicals policy at the DG Enterprise on 21.5.02; subjects at the VCI events on 16.4.02 and 13.11.02; the UBA specialised debate on 6.2.03 and the BDI specialised debate on 20.2.03 as well as the European Environmental Bureau's (EEB) conferences in Copenhagen on 27/28.9.02 and in Brussels on 31.1/1.2.03.

4 Case study analysis and development of hypotheses

As described in the previous chapter, the research process applied by the Sub-Chem project was based on an interaction between hypotheses, case studies, models and workshops.

The working hypotheses and the model were consulted in a multi-stage iterative working process, for the one part, for interpretation of the 13 case studies and, for the other part, were further developed on the basis of findings from the case studies. In addition, the contributions from the events relating to certain sectional subjects were integrated.

The procedure and the contribution of the individual elements to the project results are presented in the following section.

4.1 Evaluation and documentation of case studies

Inductive and deductive approach to the processing of case studies

Evaluation of the case studies is characterised by two courses of abstraction (abstraction or generalisation movements), which differ *only* as concerns their *ideal type* and can then be designated inductively as bottom up and deductively as top down. In the case of the 'inductive bottom up movement' the 'generation of hypotheses' is the primary objective. The aim is (as far as it is at all possible) to "let the material speak" and to draw conclusions which can be generalised from the individual experiences of the particular case studies, in most instances with the aid of a more or less refined evaluation grid.

Starting from the model and the stock of hypotheses, it was possible in the 'deductive top down movement', on the other hand, to make more accurate enquiries regarding the particular innovation processes in the case studies. As stated, both these movements must only be differentiated as ideal types.

Systematics of documentation

A structured analysis (evaluation grid) and systematic documentation were especially important for traceability of the conclusions from the case studies. Both were then used in turn as an empirical base for developing the model and for further developing the hypotheses (cf. Figure 8).

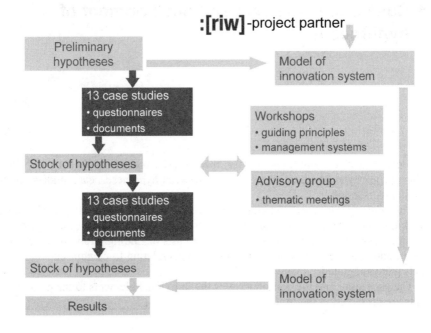

Figure 8. Operation process of model development

As part of the rough structure we examined and documented the case studies with regard to innovation level, innovation process and innovation direction in accordance with four main questions:

(a.) What does the innovation consist of? What is new? How extensive is the transformation (level of innovation[82])?

(b.) In which period, along which lines of development and with which highlights did the process take place (innovation process)?

(c.) What forces drove the innovation process on? When did each of these forces have an effect (driving forces, supporting and inhibiting factors and/or general framework conditions)?

(d.) What (potential) effects does the innovation have with regard to the environment, health, economy and social aspects (direction of innovation)?

On the basis of that, the main issue concerning the practical relevance of our examinations was: where and how can political, economic, public interest groups and other parties affect the innovation process (or change the general framework

[82] There is apparently still a complete lack of operationalisable, qualitative and/or initially quantitative indicators (e.g. in an ordinal scale) to determine the level of innovation. We made an attempt at defining these (cf. Chapters 4.3 (Hypotheses) and 5.1.1), but did not have sufficient time to pursue this matter in more detail.

conditions) in such a way that the use and especially release of hazardous substances declines for each unit of usage.

Information retrieval and data collection (as far as this was possible) was mainly carried out by means of interviews and document analyses. The search for information and the structuring of documentation for the case studies orientated around the sample questions contained in the following.

(1) Technical advantage/function: What technical (social or aesthetic) function is to be achieved (advantage of examined substance, preparation or process)? Summary of existing "devised" (as far as these are known) and "diffused" socio-technical options, which perform the desired function. Description of how the "hazardous substance" performs this function and what other aspects play a role in its application (e.g. process integration, disposal). Brief list of options to perform this function that are generally available from a technical aspect.

(2) Environment and health-related problems: What do the specific hazardous properties of the substance/product/process to be substituted (on the environment and health) entail? Who is exposed and to what extent, what effective risks exist? What measurable damage is known? What substance-specific rules exist in the field of environmental protection, occupational safety and public health? When were each of these rules introduced?

(3) Market and actors: Relevant actors (e.g. chemicals manufacturers and downstream users, enterprises, R&D, organisations, state, science) and type of market (form and intensity of competition, Fordist/quality-differentiated; B2B or B2C[83] = consumer market, supply or demand dominated)? Who is predominant (the system leader)? Who performs which role in the chain? What business volume is at stake? What are the substance quantities? What about imports/exports? What are the market shares of the various solution options? What is the relative contribution offered by the particular substance to the overall costs of carrying out the function?

(4) Innovation process: Which stages has the process gone through (time frame)?: problem recognition (other starting point), knowledge of risks, obvious impulses (initiated by which actors), highlights; present status of substitution process, significance of competing alternatives; development of legislative regulations (if they have an impulse effect), if need be windfall gains from non environment- or health-motivated innovation processes, offensive co-operations, offensive marketing.

---

[83] B2B = business to business, B2C= business to consumer

(5) <u>Innovation driving forces</u>: Type of innovation impulses (pattern of regulation, market factors/competition, public, corporate culture, strong promoters), organisational aspects (company, inter-company)? What significance did which driving force have? Did certain constellations exist on the time axis or between various actors, which accelerated or slowed down the innovation process considerably?

(6) <u>Direction of innovation (6a)</u>: To what extent did certainty of direction exist or the market actors in assessing possible solutions (assessment of possible and predominant solutions in accordance with environment- and health-related as well as economic criteria)? What information and evaluation methods were available and for which actors? What was the relation between the available knowledge and the remaining uncertainty?

<u>Direction of innovation (6b)</u>: How can the direction of innovation (health, environment and sustainability) be evaluated from the scientific viewpoint of the SubChem research group? Have, for example, the toxic risks been reduced or only been shifted?

(7) <u>Conclusion with regard to stock of hypotheses and model</u>: What are the central (prominent) features and findings of the case (especially the type of innovation, interaction of actors in the market, most important influential factors and driving forces). Which hypotheses are backed up on the basis of which aspects? Which hypotheses are refuted, if applicable? Is further information required to be able to back up individual hypotheses or grasp them more accurately? Which additional hypotheses or which changes in existing hypotheses are produced from the case? Which hypotheses are irrelevant for the case? Are there any indications of required changes in the model?

It was endeavoured to put into operation the inductive and deductive abstraction or generalisation movements mentioned at the start with the aid of these evaluation strategies in the case studies. This was done, among other reasons, in order to formulate generalisable findings (also by means of relational comparisons), which would not have been obtained solely from the inductive generalisation of the case studies. On the basis of that, at least a rough approach for the type characterisation of innovative systems was developed. In the final analysis, the aim was actually to be able to formulate (very cautious!) expectations (forecasts would certainly be too much) about the probable direction of innovation, innovation level and the diffusion rate of innovative solutions in these innovation systems on the basis of determinable system constellations and identifiable main influential factors.[84]

---

[84] Even if, from the present standpoint – i.e. after the project was completed – we had excessive demands with these further abstraction steps (regarding type characterisation and justified expectations about the 'behaviour' of innovation systems deducible from that), the project results concerning hypothesis formation, model development and recommendations for action are indeed entirely presentable.

4.2   Summary of thirteen case studies

In the 13 case studies documented processes and experiences of attempts to substitute hazardous substances in the 1980s and 1990s were examined. The aim of the choice of cases was to cover a wide spectrum of substitution conditions: consumer-close and consumer-remote products, product auxiliary materials and process auxiliary materials, SMEs and large-scale industry, environment, consumer and occupational health and safety subjects, technical and organisational innovations.

The products examined in the case studies can be grouped in accordance with the usage context of the chemicals:

- Process auxiliary agents in industrial facilities (are not included in the end products)
- Chemical products/components: used in industrial production of commodities (are included in the end products)
- Chemical products: used in crafts and by DIY users

In the following section the aim is first of all to illustrate the link between the results of the case study documentation and the hypotheses by means of a more detailed description of four selected case studies. These selected case studies represent examples from all three groups:

In the case of **metal cleaning agents** these are process auxiliary agents in industrial facilities. Industrial users and manufacturers are the main actors in the substitution process.

After use **textile auxiliary agents** remain either on the fibre (dyes, optical brighteners, finishing plasticisers) or serve exclusively as a process auxiliary material (tensides, dispersing agents) and are disposed of in waste water. This means that textile auxiliary materials are relevant for both consumer protection and also the environment. In the case of textile auxiliary agents, which at least partly remain in the product, the consumer may experience direct skin exposure. The subject of the examination is also a simple classification system developed jointly by the Association of Textile Auxiliary Agent Manufacturers and textile finishers, which facilitates the choice of more environmentally compatible substances in a pragmatic way.

Plasticisers are employed as chemicals in industrial processes, remain in the product (soft PVC) and may be released from it during use.

Chemical products used in crafts include **man-made mineral fibres**, which are used for heat and noise insulation purposes in buildings. In addition to employees in the mineral fibre industry and insulation installers, numerous other craftspeople (e.g. painters, bricklayers, roofers) and also some DIY users come into contact with artificial mineral fibres.

Representation of the individual case studies follows the documentation system described above. The hypotheses already referred to in each conclusion are presented in the following Chapter 4.3.

The higher-level synthesis of findings obtained when processing the case studies also concerning changes in the set of hypotheses and the model is presented in the summary of project results contained in Chapters 5, 6 and 7.

4.2.1   Case study "Water-based cleaning of metal surfaces"

(1) Technical advantage/function: Metal surfaces are cleaned before, between and after individual production stages either manually or automatically in order to remove greases and oils as well as other types of contamination. For this purpose petroleum-based solvents (halogenated and non-halogenated), aqueous-tensidic cleaning systems, vegetable oil esters as well as other cleaning processes (e.g. blasting or plasma cleaning) are used.

Figure 9. Ball bearings before and after cleaning (source: Co-operation Centre Hamburg)

(2) Environment and health-related problems are especially determined by health risks due to solvents, which result in damage to skin and the central nervous system (polyneuropathy, encephalopathy) in the case of chronic exposure. Some solvents are also classified as being carcinogenic (e.g. benzene, trichloroethylene) or are suspected of being carcinogenic (e.g. perchloroethylene) or reproduction-toxic (e.g. toluene).

Figure 10. Brochure by IG-Metall[85] and hazard labels

---

[85] IG Metall, Stuttgart regional directorate (publisher): Tatort Betrieb – Tückisches Gift Lösemittel, Stuttgart 1992

Volatile organic compounds (VOC) contribute to the formation of tropospheric ozone (summer smog). Certain halogenated hydrocarbons (e.g. CFCs) also destroy the stratospheric ozone layer. Chlorinated solvents are hazardous to water and, if disposed of incorrectly (e.g. burning), may emit highly toxic substances (e.g. dioxins).

But aqueous cleaning agents also contain numerous additives in addition to the actual, surface-active components (e.g. anti-corrosion agents, anti-foaming agents, biocides), which make (eco-)toxicological evaluation difficult. Basically, an important aspect of the environmental hazard of aqueous metal cleaning agents is the fact that they may contain persistent substances. In addition, surface drying requires more energy than is the case with organic solvents.

(3) Market and actors for metal cleaning agents are very varied with a heterogeneous, in some instances very split-up, user structure. A multitude of commercial small-scale users with very differing cleaning tasks and thus varied requirements on cleaning agents and also a large supply of various solutions for cleaning processes mean that the market is not transparent for users. Neither hazardous substance information, nor are the price/performance ratio of products transparent to such an extent that they could justify an informed decision to purchase.

The formulators and the trade thus play a key role in the transfer of information from the pre-supplier to the commercial end user.

At present around 65% of existing cleaning procedures in Germany function using aqueous cleaning agents, 25% using CHC cleaners and around 10% using non-halogenated hydrocarbon cleaners and other procedures.

In the **innovation processes (4)**, which ultimately caused this market ratio, two key **innovation drivers (5)** plaid a central role: a) public debates (scandals involving hazardous substances) b) state regulation (cf. Figure 11).

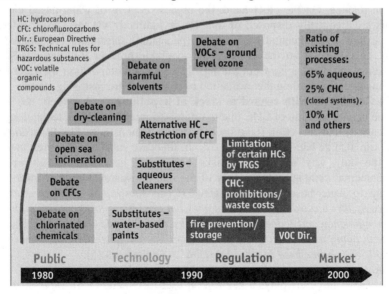

Figure 11. Innovation process for cleaning agents for metal surfaces (own diagram)

Between 1950 and 1980 traditionally used flammable "petroleum ethers" were increasingly replaced by chlorinated hydrocarbons (CHCs) for reasons for occupational safety (risk of explosion). At the same time CHCs had in many instances better cleaning properties. In the 1980s an intensive public debate surrounding "chlorinated chemicals" flared up in Germany caused i.a. by extensive ground water pollution due to CHC solvents, emissions from dry-cleaning installations in residential areas, burning of CHC waste on the North Sea and the stratospheric 'ozone hole' caused by CFCs. At the same time, an intensive discussion began in factories and unions on the subject of nerve damage caused by exposure to solvents at the workplace. A quick succession of interventions in the form of regulatory policies meant that within a period of scarcely 10 years (from 1986 to 1996) the ratio of CHC-based processes was reduced from 80% to 25%. The quantitative CHC solvent consumption even declined by more than 90% as a result of emission reduction measures.

(6a) Direction of innovation from the standpoint of market actors: Water-based cleaning agents, in comparison to CHC and aromatic cleaning agents, reduce the inhalation load at the workplace and reduce waste problems. In this area users can act relatively certain of the orientation. Possibly, however, the prejudice that "aqueous = not dangerous" led to the fact that people did not concern themselves adequately about the possible risks of aqueous cleaning agents for the environment and health. Likewise, the potential advantages of low volatile hydrocarbon cleaning agents or vegetable-oil esters were hardly perceived adequately.

(6b) Direction of innovation from the standpoint of the SubChem research group: The substitution of CHCs and hydrocarbons reduces the potential for atmospheric and ground water-related damage as well as the health burden by inhalation at the workplace. There is also no problematic waste, such as chlorinated solvent residue. On the other hand, the organic content of water-based cleaning agents are frequently disposed of directly or indirectly via the waste water. To what extent environmentally hazardous (persistent) substances are emitted can rarely be estimated on the basis of the available product information. Similar evaluation problems arise from the mostly complex composition of water-based cleaning agents regarding the evaluation of possible skin-sensitising effects.

(7) Conclusion with regard to stock of hypotheses and model: The severe decline in consumption of CHC due to optimisation of installation technology (encapsulation) and recycling (take back systems) demonstrates that hazardous solvents can also be handled with low risk. In this case substitution represented neither the only way of reducing risks nor did it necessarily result in more environmental compatibility in view of the non-transparent status of information relating to water-based cleaning agents (*"The substitution of hazardous substances by less hazardous substances does not always result in optimisation of resource efficiency and reduction of toxic risks. For example, water-based chemical products being used in an open environment do not generally entail fewer risks than oils or solvents used in closed systems"* – Hypothesis 15).

Due to the split up structure of the sector with limited know-how of downstream users, the trade plays a key role in user information (*"In small structured industries with limited user know-how the trade fundamentally affects the willing-ness for innovation of the users"* – Hypothesis 12).

Against this background the absence of practice-close evaluation systems prevents environment and health-related quality competition and favours procurement decisions taken on the basis of prejudices (*"Inadequate evaluation methods and lack of qualification of commercial users prevent an "informed" choice of product and increase the importance of emotional evaluation elements, e.g. the prejudice that "water-based = health and environment-friendly"* – Hypothesis 4).

The innovation was rather of an incremental nature, i.e. along a technology path a relatively speedy adaptation of available technologies and chemical functional principles followed on from a changed demand. The demand was initiated by a public discussion and a close succession of regulative impulses. The manufacturers of cleaning installations and aqueous cleaning agents were given the chance to increase market shares and profits (*"Higher-level sustainability innovations, e.g. the substantial reduction of resource consumption or toxic risks, can be missed out as a result of 'continuous improvement' in too small stages"* – Hypothesis 14).

4.2.2   Case study "Alternatives to reproduction-toxic plasticisers in plastics"

(1) Technical advantage/function: The case study on plastisols and soft plastic (PVC plus plasticisers) was first of all examined with the focus on car undercoating (Unterbodenschutz: UBS) in cars. To protect the car body from the impact of stones, wet and corrosion the underbody is sprayed with PVC plastisol, as PVC has the required flexible properties due to its plasticiser content of around 30%. By way of alternative, plastisols made from other plastics or underbody shells made from hard plastic can be used. The design of corrosion-free components, e.g. made of fibre composites is in the research stage at present. Soft PVC is, however, used in many consumer-close areas. Construction products such as roof tracks, flooring and vinyl wallpaper, tarpaulins for lorries, medical devices, clothing and toys likewise contain high proportions of plasticisers.

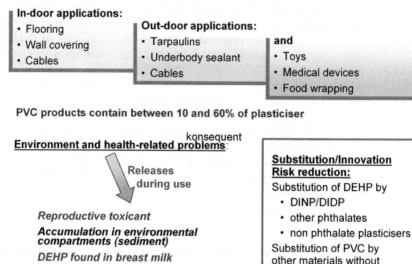

Technical advantage/function: flexible properties for polymers like PVC

In-door applications:
- Flooring
- Wall covering
- Cables

Out-door applications:
- Tarpaulins
- Underbody sealant
- Cables

and
- Toys
- Medical devices
- Food wrapping

PVC products contain between 10 and 60% of plasticiser

konsequent
Environment and health-related problems:

Releases during use

Reproductive toxicant

Accumulation in environmental compartments (sediment)

DEHP found in breast milk

Bioaccumulation in the lower food chain

Substitution/Innovation Risk reduction:
Substitution of DEHP by
- DINP/DIDP
- other phthalates
- non phthalate plasticisers

Substitution of PVC by other materials without plasticisers

Figure 12. Plasticisers in PVC products: functions and environment and health-related risks (own diagram)

(2) **Environment and health-related problems:** Risks are feared due to the use of phthalates as plasticisers, as they are not firmly bound to the PVC matrix and are released into the environment or into the air during the use and disposal stages as the result of abrasion and leaching. The prevalent plasticiser DEHP[86] is reproduction-toxic (classified as repr. Cat. II in accordance with EU Directive 67/548 since 2001). Phthalates are degraded relatively slowly in the environment with the result that especially DEHP accumulates in sediments due to the high quantities being used and the diffuse release from products. The contribution of car undercoating to the overall emissions burden of DEHP is around 1.5%. When the PVC is applied, the plasticisers contained in it may also represent a health risk for workers involved in the coating process. PVC itself is, however, also problematic for reasons relating to its disposal, as toxic chlororganic compounds (e.g. dioxins) can be formed in incomplete combustion processes. The chlorine content also restricts the choice of recycling options for the shredder light fraction (SLF) from automotive recycling. PVC plastisol may thus make it difficult to meet the recycling quotas stipulated by the Directive **2000/53**/EC on end-of life vehicles.

(3) **Market and actors:** While quite a lot of suppliers still exist on the PVC and phthalate markets, there are only three manufacturers of UBS plastisols in Germany, who are highly specialised in the manufacture of pasty preparations.

---

[86] DEHP = Di(2-ethylhexyl)phthalate

These companies also develop plastisols on the basis of other plastics, which however frequently also contain plasticisers. Underbody hard shells, on the other hand, are offered by other manufacturers. The market price for PVC plasticiser systems as a whole is characterised by the fact that the polymer (PVC) and the principal plasticiser (DEHP) are produced in very high volumes (in the EU 15 m. tons of PVC are produced per annum, with 600,000 tons of DEHP being produced in 1997) and can be combined flexibly in order to obtain certain technical product qualities. The longer-chain phthalates DINP and DIDP[87] are also high-volume chemicals (> 100,000 t/a). Soft PVC products are thus generally incomparably cheap.

(4) Two substitution stages can be identified for the **innovation process** in the case of car undercoating: after being classified by the substance manufacturers (in 1994) as reproduction-toxic Cat. III, DEHP was included in the VDA list[88] of declarable materials (1995), according to which suppliers are obliged to declare products containing DEHP accordingly. Plastisol manufacturers replaced DEHP within a few months by a mixture of DINP and DIDP, although at that time it was uncertain as to whether these phthalates did not have similar toxicological properties.

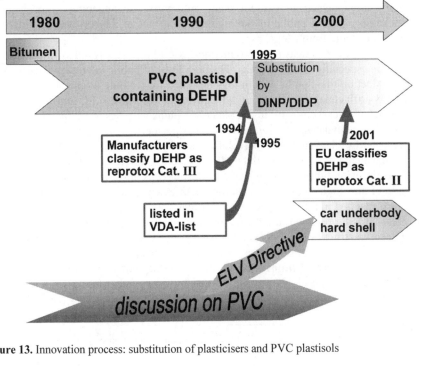

Figure 13. Innovation process: substitution of plasticisers and PVC plastisols

---

[87] DINP = diisononyl phthalate, DIDP = diisodecyl phthalate
[88] List of declarable materials in automobile manufacturing – Substances in components and construction materials (previously VDA List 232-101), http://www.mdsystem.com/html/de/home_de.htm

Pressure for substitution also existed for the material PVC due to the public PVC discussion and also indirectly due to the DIRECTIVE 2000/53/EC on end-of life vehicles. Dismantlable underbody hard shells made of polypropylene are suitable to solve these problems. However, the design of the underbody has to be adapted to the use of a hard shell, which can ultimately only be achieved in combination with a change of model.

(5) Innovation drivers: With the legal classification of DEHP in 2001, the substance had to be labelled with a skull pictogram throughout Europe (as well as displaying the R-phrases R 60 - 61[89]). Following this, DEHP was also substituted in many areas by the phthalates DINP and DIDP for reasons of occupational health and safety. There is no obligation to label DINP and DIDP yet. On the European plasticiser market this caused a significant shift. The DEHP ratio thus only accounted for 25% of the total plasticiser market in 2003 (compared with 42% in 1999). The ratio of DINP/DIDP rose to 55% during the same period (from 35% in 1999).

The PVC discussion as a driving force for innovation affected in particular consumer-close products. Alternative materials did not gain the upper hand for any of the mass products, however "PVC-free" as a product strategy was considered and partly implemented at least for many automotive manufacturers as well as for packaging, even if public marketing (e.g. by means of a label) was considered to be problematic from a legal aspect[90].

The aspect of releasing reproduction-toxic plasticisers resulted in substitution only in especially sensitive areas. The high level of public interest in the release of reproduction-toxic phthalates in toys designed for teething infants and babies was highly influential. Not least of all due to an emergency EU ordinance, which banned the marketing of toys containing phthalates designed for chewing and suckling, phthalates were replaced by other plasticisers and furthermore PVC was replaced by plasticiser-free polyolefins or ethylene-vinylacetate. Another controversial factor was the use of DEHP in medical devices (such as blood storage bags, infusion tubes), from which the plasticiser is absorbed by patients. As a result of the high application-specific requirements, substitution is considerably more exacting in this case. For these particularly sensitive and relatively expensive areas, BASF presented in 2002 the plasticiser DINCH[91], which was developed for a specific target (since 1997) and subjected to extensive toxicological testing and for which no CMR or other long-term effects were detected.

(6a) Direction of innovation from the standpoint of market actors: When DEHP in UBS plastisols was substituted by other phthalates in 1995 due to the manufacturers' classification, the users (plastisol manufacturers and automotive

---

[89] R 60: may impair fertility.; R 61: may cause harm to the unborn child (Council Directive 67/548/EEC)

[90] The question to what extent the objectively accurate statement "PVC-free" associates general environmental friendliness in the case of a product and is thus misleading is dealt with by a court judgment (Federal Supreme Court: I ZR 76/94), which ruled that this statement was permissible.

[91] DINCH = diisononyl cyclohexane dicarboxylate (also diisononyl hexahydropthalate).

manufacturers) did not have adequate data to evaluate the substances. As DINP and DIDP are only slightly different from DEHP from a chemical aspect, at least a similar potential effect had to be expected. Nevertheless, the plastisol manufacturers did employ these substances, as no low-cost alternative with similar technical properties was available.

Concerning the evaluation of PVC, users were repeatedly confronted with contradictory information. Comparisons in life-cycle-analyses (which in the majority of the cases exclude toxicological effects) frequently speak in favour of PVC, which does come off better because of its lower resource requirements. The public discussion about phthalates in children's toys, the problems involved in the disposal of PVC and the consequences of large-scale fires in structures containing PVCs (such as the one at Düsseldorf airport) however, lead to a scandalisation both for PVC and also for phthalates.

From a user's point of view, one advantage of using underbody hard shells is the saving in weight (strategies to reduce fleet consumption of petrol have in the meantime become part of the environmental policy of many companies), improved recycling options, fast cycle times for assembly and the reduction of overhead work. This is counterbalanced by problems in the area of corrosion protection (accumulation of damp) and higher material costs. The processes to be taken into consideration in order to reach a decision here also increasingly depend on non-scientific factors (environmental policy, guiding principles, brand image).

(6b) Direction of innovation from the standpoint of the SubChem research group: Substitution of the reproduction-toxic plasticiser DEHP for DINP and DIDP may be considered to be an improvement from a health aspect, as the EU risk assessments[92] for these substances that have now been published do not identify any comparable potential for health risks[93]. For the assessment of alternative plasticisers (e.g. citrates, adipates or alkylsulphonic acid esters), which replace e.g. DINP in toys, the data pool was inadequate at the time of substitution, with the result that it could not be ruled out that new health risks would be created by these substances[94]. Potentially harmfull exposure, however, also depends especially on the mobility of these substances in plastics. Knowledge about the release mechanisms of additives is as yet very limited.

All three phthalates degrade considerably worse under environmental conditions than they do in laboratory tests, however they cannot be classified on the whole as being persistent. The substitution of DEHP by DINP/DIDP in plastics for

---

[92] EU risk assessment for DINP (5/01) and DIDP (5/01), as reported in France.

[93] Merely one restriction was made relating to DIDP as the main plasticiser used in children's toys.

[94] Scientific Committee on Toxicity, Ecotoxicity and the Environment (CSTEE): Opinion on toxicological characteristics and risks of certain citrates and adipates used as a substitute for phthalates as plasticisers in certain soft PVC products, September 1999. A risk assessment has now been compiled for the widely used citrate plasticiser ATBC. The CSTEE comes to the conclusion that there are no safety concerns relating to the use of ATBC (o-acetyl tributyl citrate) in toys designed for chewing (CSTEE 41st plenary meeting, 8th January 2004).

external applications (such as car undercoating) only represents an incremental improvement, as substances are still being emitted into the environment. Due to the very substantial volumes being used[95] and the continuous leaching from existing products already described, an environmental concentration is formed at a relatively high level even without persistence (water > 1 µg/l, sediment > 10 mg/kg)[96], which can be referred to as "quasi-persistent". In addition bioaccumulation takes place, at least in the lower levels of the food chain. In general environmental accumulation is increasingly being recognised as an (environment hygiene) problem, even if no problematic effects are detected in the ecosystem caused by the substance (such as e.g. reproduction-toxic effects).

By way of alternative to UBS plastisols, an underbody hard shell not only reduces the release of plasticisers during the usage phase. Due to reduced weight and reduced air resistance (CW value) fuel consumption of vehicles is also reduced, which has a positive effect as part of a life-cycle assessment. After dismantling the plastic part can be recycled separately.

(7) Conclusion with regard to stock of hypotheses and model: In the present case we can observe various substitution stages associated with different driving forces. The substitution of DEHP in toys or medical devices can be put down to pressure exerted by public interest, caused by health risks for sensitive consumer groups. DEHP in car undercoating is rather an occupational health issue. The release of large quantities of phthalates into the environment, which are not easily degradable, was not an explicit reason for action either for the manufacturers or for the users.

Both PVC and also phthalates were 'scandalised' in public discussion, with the result that in certain individual cases substitution no longer was based on scientific risk analysis alone (Hypothesis 4 cf. above).

PVC-DEHP material systems are incomparably cheap due to their high market volumes. At the same time, DEHP is one of the best examined substances on the European market from a toxicological aspect, although no greater assessment safety has been created as a result of this in the past 25 years[97] (*"The demand for 'toxicological certainties' consolidates a lock-in in the case of high-volume existing substances"* – Hypothesis 3).

Establishing alternative plasticisers and material systems on the market is correspondingly difficult. This is especially the case for "new substances" (hypothesis 2). These competitive advantages (high volumes, low price, extensive examined existing substances) can only be overcome by companies with considerable capital, who manufacture DEHP themselves (*"The existing chemicals legislation*

---

[95] In total more than 900,000 tons of phthalates are processed in Europe every year. The volume of DEHP used (currently barely 30%) is declining in favour of DIDP/DINP (almost 60%) (source: ECPI 2003).

[96] EU risk assessment (draft) for DEHP 2003.

[97] In addition to contradictory test results, the remaining uncertainties concerning the transferability of animal testing to humans still leave significant scope for interpretation with regard to the relevance of the effects observed and the safety factors to be deduced from them.

gives existing substances certain competitive advantages, which hamper innovation" – Hypothesis 1).

BASF is thus currently endeavouring to establish the substance DINCH as a new registered substitute (entirely tested plasticiser without any hazardous properties) for DEHP in high-price market segments (e.g. medical devices). Soft PVC users substitute on the material level, if they can achieve further process-related and/or quality-related optimisations in the course of technological development (e.g. underbody hard shells) (*"Environment and health-compatible substance properties are additional qualities in business-to-business (B2B) markets. They are of relevance almost only for manufacturers, who market their products on demand-dominated, saturated markets with differentiated quality production"* – Hypothesis 8).

4.2.3 Case study "Biosoluble, man-made mineral fibres for insulation in structures"

(1) Technical advantage/function: The mineral wools (MW) belonging to man-made mineral fibres are used in the construction industry for heat and noise insulation purposes in structural engineering. A common feature of all insulation materials is the fact that they have a large volume with low weight due to the many small hollow spaces that they contain. Besides mineral wools, hard foam (e.g. polystyrene, polyurethane) as well as cellulose, cork and lightweight building boards are used.

Figure 14. Various insulation materials made of mineral wool (source: Bau-BG, Frankfurt)

The technical suitability of insulation materials is described by a series of technical parameters relating to the material (heat conductivity and heat storage capacities, damp protection, fire protection class, noise-insulating effect, properties related to building biology, e.g. content of hazardous substances such as flame retardants and insecticides). Mineral wools have certain advantages over other insulating materials, especially in the area of fire protection and the material price[98] and also due to their simple processing.

(2) Environment and health-related problems are dominated by risks emanating from fibre dusts, which are released during the production and processing of mineral wool products and in refurbishment and/or demolition work. If the fibres are sufficiently small and are inhaled, they can enter the lung. They can also induce skin and respiratory diseases and some are also potentially carcinogenic.

The carcinogenic effect of a fibre depends on the concentration (exposure[99]), the geometry of the fibre, which is also called fibre dimension (which also determines the inhalability[100]) and also the biodurability (persistence in the body or dissolution of the fibres by the body).

With regard to biodurability the resistance of the fibres to elimination mechanisms in the lung is decisive from a toxicological aspect. Biosolubility is an essential parameter for this.

It is evident that fibre geometry and biodurability are the decisive criteria for the carcinogenic properties of fibres. However, there are differing opinions between fibre manufacturers and occupational health and safety experts in the EU and Germany concerning the examination and measuring methods and the evaluation of the carcinogenic properties.

(3) Market and actors: The market of mineral wool manufacturers comprises a few, mostly medium-sized enterprises (approximately 5 companies with around 20 plants in Germany). They are dominated by two well known, also globally active companies. Due to the high volume of the commodity, the market is essentially organised on a regional to national basis. The manufacturers are faced with employees in the insulation trade, who are well organised in unions. The mineral wool industry in Germany employs around 5000 employees in the manufacturing and industrial pre-packaging sectors. In the structural engineering, civil engineering and technical insulation industries approximately 30,000 insulation workers handle man-made mineral fibre products. The number of employees in other areas and in other parts of the construction industry, who handle man-made mineral fibres at least on a sporadic basis (e.g. bricklayers, dry construction workers, painters, carpenters, roofers, assistants i.a.) is estimated to be 450,000. In the 1990s the production of insulation materials rose substantially. In 1999 approximately 33 m. m^3 of insulation materials were manufactured. Of this figure mineral wools account for a market share of 58%. The mineral fibre industry accounts for 12.7% of the entire turnover of the German glass and mineral fibre industry of DM 16,600 m., i.e. a turnover of DM 2,100 m.

---

[98] Prices for a k-value (thermal transfer coefficient) of 0.2 in €/m²: Mineral wool 5- 17.5; polystyrene 10; polyurethane 15 -20; cork 50; wood wool-lightweight building board 100, cellulose 12.5-20 (loose) or 20 (plates) (ENRW 01).

[99] The Geramn air threshold value (technical reference concentration TRK) for biopersistent mineral wools and ceramic fibres 250,000 f/m^3 (certain areas 500,000 f/m$^{3)}$ (German TRGS 900/901).

[100] Fibres with a critical fibre geometry (or WHO fibres) are considered to be those with a length > 5 μm, a diameter < 3 μm and a length-to-diameter ratio < 3 to 1.

(4) Innovation process: Mineral wools have been used as the most important substitutes for asbestos and then themselves came under suspicion of causing cancer. Along with the realisation that fibres with a certain chemical composition can be dissolved in the lung, a new quality – biosolubility – was developed for man-made mineral fibres. Biosoluble fibres were marketed for the first time in the mid 1990s, when the pressure on existing man-made mineral fibre products increased. Besides an imminent ban on biodurable fibres in the construction industry as well as the demands of professional users (insulation workers) and their representatives, the series of regulations relating to the classification of fibres grew:

Regulation of man-made mineral fibres was concerned with two formally separate areas:

– classification: EU Directive 97/69/EC, §4 German Ordinance on Hazardous Substances
– occupational safety: German TRGS 521, German Ordinance on Hazardous Substances Annexe V no. 7, German ChemVV

EU classification: The basic classification as carcinogenic fibres depends on the content of alkaline and alkaline earth metal oxides (Na_2O, K_2O, CaO, MgO, BaO) (EU Directive 97/69/EC). In accordance with this directive, mineral wools are defined as "man-made glass (silicate) fibres with random orientation". With alkaline oxide and alkali earth oxide content in excess of 18% by weight they are classified as carcinogenic Cat. 3 "Suspected carcinogenic effect"[101] and also "irritating". Under certain conditions, as stated in Nota Q of the Directive ("Restriction of liability"/"Opting out criteria") classification as carcinogenic is not compulsory:

Classification for Germany: Man-made mineral fibres were included in Annexe V no. 7 of the German Ordinance on Hazardous Substances in 1998. Criteria for restriction of liability were set out here. The evaluation methods are stated in TRGS 905 "Index of carcinogenic, mutagenic or reproduction toxic substances" no. 2.3 (German Federal Ministry of Labour announcement in accordance with § 52 Para. 2 German Ordinance on Hazardous Substances). The German Ordinance on Hazardous Substances and the German ChemVV were amended on 01.06.2000. In accordance with this amendment, the manufacture, marketing and use of new products containing mineral wools with biopersistent fibres is prohibited for heat and noise insulation in structural engineering purposes in Germany (Annexe. I no. 22 German Ordinance on Hazardous Substances and the German ChemVV).

---

[101] "Substances, which give rise to concerns due to possible carcinogenic effects on humans, but which cannot be evaluated definitively due to insufficient information. Knowledge exists on the basis of suitable animal testing, but this is not sufficient to classify the substance in Category 2."

	Basis			K2	K3	K0	EU	D
1	chemical composition	Index of carcinogenicity (KI)		≤ 30	30 ... 40	≥ 40 or cond. 2,3	-	x
		alkali and earth alkali oxides		≤ 18 %	> 18 %	cond. 2,3,4	x	-
2	B i o p e r s i s t e n c e	intraperitoneal-Test		**carcinogenicity**			x	x
				high	medium	no indication		
3		intratracheal test		**half life (days)**				x
				> 40		≤ 40		
				≥ 40		< 40	x	
4		inhalation test	short-term	**half life (days)**				-
				≥ 10		< 10	x	
			long-term	relative pathogenicity or neopl. changes		absence of pathogenicity and changes	x	-
5		In-vitro-testing		**elimination rate**			-	-
				low	medium	high		(in preparation)

Figure 15. Classification of man-made mineral fibres[102]

Occupational health and safety: The Ordinance on Hazardous Substances Annexe V no. 7) and the TRGS 521, which was last updated in May 2002, govern the handling of mineral wool fibres in Germany. When processing biosoluble mineral wool products, merely the minimum measures for the protection of workers against dust as per TRGS 521 no. 4 must be observed. Since the 2000 ban the handling of biopersistent mineral wools, which may release fibre dusts suspected of being carcinogenic (no. 2.3 TRGS 905), is only possible or permissible during demolition, refurbishment, maintenance and repair work. Special protective measures are required for this (TRGS 521 no. 5). The German Ordinance on Hazardous Substances generally sets out the obligation to find a substitute as well as the duty to give notice as well as prescribing protective measures. This means that the innovation was driven on by the set of regulations together with a fierce dispute concerning the parameters of hazardous substance classification.

In September 1998 the RAL Quality Community on mineral wool, which represents 90% of mineral wool manufacturers, was established on the basis of an sector-wide agreement. It is also the owner of the quality label "mineral wool products" and works to promote the diffusion of biosoluble mineral wools (GGM 01). The RAL label on the products signifies easily identifiable information, which at the same time has been verified and continuously examined by a neutral institution, as to whether a mineral wool product is released as per Annexe V no. 7.1 of the German Ordinance on Hazardous Substances.

---

[102] Löffler and Reuchlein: Sicherer Umgang mit künstlichen Mineralfasern, in: Arbeitsmed. Sozialmed. Umweltmed. 35, 12, 2000

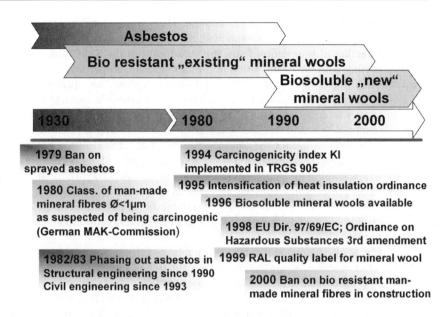

Figure 16. Innovation process in the case of man-made mineral fibres for heat insulation in structures (own diagram)

(5) Innovation drivers: Scientific findings concerning the effect mechanisms of mineral fibres (fibre dimension, biosolubility) and experience with the carcinogenic effects of asbestos have created certain impulses for the whole innovation process. Some health-conscious end consumers, but especially commercial users, have been sensitised accordingly. The case of asbestos is one example as to how scientific findings, on the one hand, can play an important role in promoting substitution, but on the other hand how insisting on 'complete findings' or instrumentalising gaps in findings can also delay or block substitution processes. The latter applies e.g. for the highly controversial discussion surrounding classification criteria and the transferability of animal testing, which are ultimately also reflected in the differences between regulations on both national and European levels.

(6a) Direction of innovation from the standpoint of market actors: It has been confirmed that the strategy for the substitution of asbestos by mineral wools could not be certain in its direction in retrospect because certain evaluation criteria were not applied (critical fibre dimension) or were missing (biosolubility) at the time of substitution. A central aspect of this case is the controversially discussed (and partly also legally contested) evaluation problem of health hazards and the resulting legal classification.

By force of circumstances the manufacturers of mineral wools participated in the development of test methods to distinguish between hazardous and non-critical fibres as well as appropriate criteria for legal classification. A delay in the innovation process was caused by the discussion as to which criteria for the restriction of liability should actually be legally permissible. There was and still is general scepticism on the part of the manufacturers with regard to national regulations.

(6b) Direction of innovation from the standpoint of the SubChem research group: Starting with the examination of the carcinogenic properties of chemically inert asbestos fibres, the significance of the fibre dimension was discovered first. So similarly, a hazard by accordingly dimensioned man-made mineral fibres (or WHO fibres) was deduced. Animal and laboratory testing demonstrated that the chemical composition and the biopersistent/biodurable properties determined by them are also responsible for the carcinogenic effect of mineral fibres. On the basis of biosolubility the 'internal' exposure to carcinogenic fibres is reduced considerably, thus making a significant contribution to occupational health and safety and consumer protection. Reducing the hazards by means of a relatively simple principle, i.e. the shift to manufacturing biosoluble fibres (alteration of the chemical composition of fibres without any major technical and organisational changes), is a peculiarity.

The aspects of occupational safety and public health are seen via-à-vis the arguments of environmental protection. Heat insulation of buildings is the most important measure to reduce private energy consumption. The German Heat Insulation Ordinance[103] and the German Energy Conservation Ordinance which has been in force since February 2002 for new and old buildings alike increase the requirements for heat insulation. So far mineral fibre products are most frequently used for insulation purposes due to their easy workability, their good fire prevention and damp properties and also due to their favourable price in comparison with other insulation materials. There was thus a conflict of aims between environmental protection (energy conservation = resource preservation and reduction of CO_2 emissions (greenhouse effect)) and public health (release of potentially hazardous fibres).

(7) Conclusion with regard to stock of hypotheses and model: Against the background of experiences with asbestos, the fibre discussion met with a relatively large response in public and thus ultimately motivated the manufacturers to market biosoluble products (Hypothesis 4 cf. above).

The competition between two major manufacturers of mineral wools, especially relating to the measuring methods for the criterion of "biosolubility", resulted in a delay to the introduction of biosoluble mineral wools, i.a. because the official bodies came under pressure to find a justification for the measures of hazardousness leading to market restrictions. New developments were delayed as long as possible until no other action was possible. Knowledge of forthcoming classifications was thus used as a competitive advantage, on the one hand. On the other hand, the adaptation of more binding regulations to the actual state-of-the-art was also delayed by this. (*"The case shows the limits of possibilities for state intervention and the lengthy periods until regulative impulses become effective."* – Hypothesis 6 "Regulative drivers").

---

[103] The basis of the German Heat Insulation Ordinance is the German Energy conservation Act dated 1976. The first Heat Insulation Ordinance came into force in 1977, with amendments in 1982 and 1994.

In the final analysis, increasing regulative pressure and the heightened attention of the public has meant that biosoluble fibres have been manufactured since the mid 1990s. A co-operation to promote the diffusion of biosoluble mineral wools was established with the RAL Quality Community on mineral wool established in 1998, which represents 90% of mineral wool manufacturers. (*"Co-operation networks and shared guiding principles may be an important prerequisite for the successes of innovation"* – Hypothesis 10 "Co-operation networks").

The peculiarities of this case are to be found (a) in the type of substitution – the development of alternative fibres without[104] any carcinogenic potential (biosolubility) and (b) in the fact that substitution has been implemented 100% at least in Germany.

4.2.4   Brief presentation of other case studies

Presentation of the individual case studies follows the above described documentation system. The hypotheses referred to in the particular conclusion are described in detail in Chapter 4.3.

For detailed information the complete versions of the case studies are available from the authors.

Case study "Biocide-free cooling lubricants and minimum quantity cooling lubrication"

(1) Technical advantage/function: Cooling lubricants are used in large quantities as manufacturing auxiliary material for almost all machining operations. The main function of cooling lubricants is to cool and lubricate the part being machined and also to rinse away the turnings. Minimum quantity cooling lubrication and dry machining are used as alternatives for flooding cooling lubrication (wet machining).

(2) Environment and health-related problems: Skin diseases, caused by dampness, alkalinity of cooling lubricants and skin-hazardous additives (e.g. sensitising emulsifying agents, corrosion inhibitors, biocides) are the primary health hazards caused by water-soluble cooling lubricants. Individual substances may be contained in the emulsion being used, which are capable of causing respiratory problems (e.g. mould, endotoxins, amines) as well as cancer (e.g. N-nitrosamines, PAH).

(3) Market and actors: The multitude of commercial users (approx. 100,000 companies) with different machining tasks and thus varied demands on cooling lubricants and also a wide range of products on offer cause the market to be non-transparent for the users. There are around 80 cooling lubricant suppliers in Germany (ranging from international mineral oil groups to SMEs).

---

[104] In accordance with the present state-of-the-art.

(4) <u>Innovation process</u>: Cooling lubricants on the basis of vegetable or animal oils have been used for centuries. In the 1950s emulsions came onto the market, which due to the better physical cooling effect of the water met the increased demands for faster machining processes and higher quantities. Substitution processes for cooling lubricants have so far only been applicable to individual ingredients (PAH, formaldehyde, nitrite, boric acid) and technical optimisation. Due to the effort spent on maintenance and the increased costs of waste reduction, dry machining and minimum quantity cooling lubrication have been further developed since the late 1980s.

(5) <u>Innovation drivers</u>: Increasing cost pressure as the result of disposal costs and high expenditure on monitoring, maintenance and care of cooling lubricants primarily promoted waste-reducing processes such as minimum quantity cooling lubrication and dry machining. With the action "Poison cocktail cooling lubricant" (1989) IG-Metall sensitised affected employees. Results included, among other things, the publication of new regulations by the employers' liability insurance (BGR 143) and in 1993 enactment of TRGS 611. In the meantime, the lubricant manufacturers' associations, the employers' liability insurance and IG-Metall have compiled a list of substances, that is constantly updated, which should not be contained in cooling lubricants[105].

(6) <u>Direction of innovation (from the standpoint of market actors – 6a)</u>: The choice of the right cooling lubricant is difficult due to the multitude of products and ingredients. In many instances hazards occur only as the result of contaminations or other alterations to the cooling lubricant. Users rely on information relating to properties and handling (maintenance and care) by the manufacturers.

<u>Direction of innovation (from the standpoint of the SubChem research group – 6b)</u>: By substituting cooling lubricants by dry machining and/or minimum quantity cooling lubrication, a reduction of hazards caused by cooling lubricants (skin and respiratory deseases) can be noted. Furthermore the risks are reduced caused by insufficient or incorrect handling of cooling lubricants.

(7) <u>Conclusion with regard to stock of hypotheses and model</u>: Due to the complex multitude of products, the trade and distribution may affect the choice of alternatives in each case by means of user information. ("Influence of trade and distribution" – Hypothesis 12). The high disposal costs exercise innovative pressure to reduce volumes of cooling lubricants (Hypothesis "Cost internalisation"). Some ingredients of cooling lubricants were the subject of public discussions (e.g. formaldehyde, chlorinated paraffins) ("Emotional drivers" – Hypothesis 4).

---

[105] VKIS-VSI-IGM

Case study "Biodegradable mould release agents made of natural raw material "

(1) <u>Technical advantage/function</u>: Before casting, mould release agents are applied to the mould surface/formwork. They prevent the concrete from adhering to the formwork after hardening with the result that the concrete surface is damaged while it is being removed. A distinction is made for the application in pre-cast concrete part plants and on site (site-mixed concrete). The former have higher demands on the technological effectiveness of the mould release agents. Vegetable-oil-based mould release agents (aqueous ester emulsions) are an alternative to mineral-oil-based mould release agents, some of which contain solvents.

(2) <u>Environment and health-related problems</u>: The main problem is the environmentally open application of conventional mineral-oil-based mould release agents. Mineral oils are persistent in the environment, mobile in soils and may contaminate the ground water. They have a highly toxic effect on water and soil organisms. Solvent contents (if used for pre-cast part plants) cause VOC emissions. Health risks (such as lung, skin and nerve damage) may result for users due to mould release agents. In pre-cast part plants the danger of fire and explosion is also present.

(3) <u>Market and actors</u>: In the market for mould release agents a distinction is made between a quality-differentiated market in the area of pre-cast part plants and fairfaced concrete and a heavily cost-oriented competition in site-mixed concrete. The manufacturers are for the one part mostly medium-sized manufacturers of construction chemicals, and for the other part mineral oil groups. Solvent-free, mineral-oil-based mould release agents are prevalent. The proportion of products containing solvents is estimated to be approximately 20%, the proportion of vegetable-oil-based mould release agents is around 5%[106].

(4) <u>Innovation process</u>: Besides improving the technological properties, the aim of developing mould release agents since the 1980s has been to reduce problematic substances and increase biodegradability. Measures included cleaning the basic oils contained in mould release agents (extraction process), since 1992 use of easily biodegradable vegetable-oil-based mould release agents and increased degradability of mineral-oil-based mould release agents Due to the higher price[107] and a more complex application method; the vegetable-oil-based mould release agents are mainly used in pre-cast part plants.

---

[106] These figures refer to the German market.

[107] 2-3 times more expensive than site-mixed concrete-concrete separating agents, approx. 30% more expensive than high-quality, solvent-based concrete separating agents for pre-cast part plants.

(5) Innovation drivers: The initiator for development of vegetable-oil-based mould release agents came from The Netherlands and Switzerland in the mid 1990s. It was instigated by high disposal costs and the costs for cleaning up the ground after contamination by mould release agents. These experiences from abroad motivated manufacturers in Germany to develop vegetable-oil-based mould release agents.

(6) Direction of innovation (from the standpoint of market actors – 6a): The primary concerns are the quality and price of products. As possible damage to the concrete surface due to incorrect usage of mould release agents causes very high costs for repairs, new products with another application technology are considered with scepticism.

Direction of innovation (from the standpoint of the SubChem research group – 6b): The advantage of vegetable-oil-based mould release agents is, on the one hand, in their character as renewable raw materials, their biodegradability and on the other hand the reduction of VOC emissions as well as risks of fire and explosion.

(7) Conclusion with regard to stock of hypotheses and model: In the case of site-mixed concrete costs are the decisive factor. The higher costs for vegetable-oil-based mould release agents are thus a hindrance to substitution ("Fordist market" – Hypothesis 8). As a result of the competition of vegetable-oil-based mould release agents, conventional mineral-oil-based mould release agents were also further developed in relation to their biodegradability ("Cross-adaptation" – Hypothesis 5).

Case study "Biosoluble, man-made mineral fibres in automotive silencers"

(1) Technical advantage/function: Basalt rock wool is used for noise insulation in automotive exhaust units for the purposes of noise absorption. Vital technical requirements are temperature resistance, noise absorption behaviour, corrosion resistance and mechanical stability. Now textile continuous glass fibres and biosoluble mineral wools are also used for this application in exhaust units.

(2) Environment and health-related problems: In production and processing (packaging, installation of basalt rock wools) fibre dust is released. This dust may cause skin and respiratory diseases. Basalt rock wools are biopersistent mineral wools and, as such, have a certain carcinogenic potential. Conversely, textile continuous glass fibres are not considered to be carcinogenic due to their structure (fibre diameter = 24 µm) and biosoluble mineral wools due to their solubility in the lung.

(3) Market and actors: A few fibre manufacturers and customers (suppliers of the automotive industry, as well as the automotive manufacturers themselves) are at the market. Between the fibre manufacturers and the silencer manufacturers a multitude of intermediate sub-contractors are involved, who purchase fibre

materials, partly package them and then resell them. At present the market share of basalt rock wools[108] is declining significantly and the market share of textile continuous fibres and biosoluble mineral wools is growing.

(4) Innovation process: The development of biosoluble fibres for use in silencers was a difficult task from a technical aspect. On the one hand, the materials had to be soluble in the lung environment, on the other hand, they had to guarantee durability in the acidic environment of the exhaust flow. The change-over period until the release of "new" materials in the automotive industry is no shorter than 2-3 years depending on production conditions (investment and test cycles). Since the start of 1999 biosoluble fibres have been produced in series for use in silencers.

(5) Innovation drivers: Artificial mineral fibres were included in Annexe V of the German Ordinance on Hazardous Substances in 1998. In 2000 biopersistent mineral fibres were banned from use in construction. These regulations, together with the in some cases public discussion of fibres, also provided impulses for substitution by biosoluble fibre materials in silencers.

(6) Direction of innovation (from the standpoint of market actors – 6a): A complex assessment situation (classification, criteria for restriction of liability, transitional regulations[109]) resulted in confusion with regard to the need for substitution for manufacturers and users of basalt rock wools.

Direction of innovation (from the standpoint of the SubChem research group – 6b): If biopersistent basalt rock wools are replaced by biosoluble mineral fibres (in manufacture, processing, recycling), the 'internal' exposure that causes cancer is reduced. This makes a significant contribution to occupational health and safety.

(7) Conclusion with regard to stock of hypotheses and model: The substitution process in the construction industry had the effect of providing impulses ("Cross-adaptation" – Hypothesis 5). Automotive manufacturers as brand manufacturers are especially vulnerable to scandals. Concerning carcinogenic fibres it must be assumed that the public and therefore also car buyers were highly sensitised due to the asbestos problem ("Emotional drivers" – Hypothesis 4).

---

[108] Up until approximately 1998 almost 100 % of filling materials used in silencers still comprised basalt rock wools.

[109] In accordance with the transitional regulations in § 54 Para. 4 of the German Ordinance on Hazardous Substances biopersistent mineral fibres for use in automotive silencer units are exempted from the substitution obligation in accordance with Annexe V no. 7.2. and the duty of disclosure in accordance with no. 7.3. until October 2003. Concerning the use of fibres in vehicles, this means that an already existing substitute does not have to be used until that time.

Case study "Man-made mineral fibres in automotive catalytic converters"

(1) Technical advantage/function: Ceramic fibres are used in automotive catalytic converters as bearing and adjustment materials for the catalytic converter (monolith), where the chemical reactions for exhaust cleaning take place. They are also used for thermal and acoustic insulation. Series-tested ceramic fibre substitutes for converter-specific usage conditions are not yet available[110].

(2) Environment and health-related problems: Fibre dusts are released during manufacturing, packaging and recycling of catalytic converters. They can cause skin and respiratory diseases, some of them have carcinogenic potential. The carcinogenic effect of ceramic fibres is consistently assessed both in accordance with EU legislation and German regulations as K2 (suspected of having a carcinogenic effect in humans).

(3) Market and actors: In Germany there is one manufacturer of suitable ceramic fibres (three in Europe and two in Japan). Workers in vacuum forming, modular structure and punching plants as well as other plants, where ceramic fibre products are installed and used, also handle ceramic fibres.

(4) Innovation process: Unlike silencers, a substitute for ceramic fibres in catalytic converters is not yet available because very stringent technical requirements have to be met. The test cycles and clarification of the technical and other properties (occupational health and safety and environmental protection) are also correspondingly extensive. Solutions are being developed in cooperation with manufacturers, users and official bodies.

(5) Innovation drivers: An important impulse was the classification of ceramic fibres as carcinogenic category K2 (Directive 97/69/EC), which has been applicable since January 1998. Since July 2001 there has been a ban on the marketing of ceramic fibres for the "general public". An impending ban also for the industrial sector caused alarm in the automotive industry. A scientific study by Wuppertal University[111] also directed attention on the subject of "ceramic fibres in catalytic converter recycling".

(6) Direction of innovation (from the standpoint of market actors – 6a): For automotive and catalytic converter manufacturers the direction was clearly "away from ceramic fibres classified as K2" as per the 1998 EU classification. Official bodies (especially FIOSH) were consulted for assessing the hazardousness and were thus included in developing substitutes. A development was thus accelerated with the highest possible security of direction.

---

[110] This was the case at the time of data collection (06/2003); as of 2004 substitutes are to be available and are also to be used in the automotive industry.

[111] Kahl-Mentschel, A. (1999)

Direction of innovation (from the standpoint of the SubChem research group
– 6b): The substitution of ceramic fibres in the construction of catalytic converters would avoid exposure (manufacture and recycling) to potentially carcinogenic fibres. Inclusion of specialised official bodies in the evaluation of
alternative materials is one step towards greater security of direction.

(7) Conclusion with regard to stock of hypotheses and model: Due to the "vulnerability" of the automotive industry to public scandals the subject of "carcinogenic fibres in catalytic converters" was not discussed openly by the automotive industry ("Emotional drivers" – Hypothesis 4). The creation of a "round
table" comprising official bodies, automotive manufacturers, catalytic converter manufacturers, filling material manufacturers and scientists can be considered as an instrument supporting innovation. (Hypothesis 10 – "Co-
operation networks").

Case study "Low-solvent automotive series coatings"

(1) Technical advantage/function: Automotive coatings not only have to cope
with the heavy stresses of car bodies (impact of stones, corrosion, UV) and
optical demands of users (colour, sheen, durability), but also facilitate speedy
and reliable processing in series manufacture. Various coating layers are
combined (primer, filling, coloured coating and clear coating) in automotive
series coating.

(2) Environment and health-related problems: Conventional coatings contain solvents, which contribute to health risks as well as the formation of tropospheric
ozone (summer smog), if they are not eliminated by elaborate emissions
cleaning processes. In the coating process large quantities of energy and water
are also consumed and excess coating sludge has to be disposed of.

(3) Market and actors: In automotive series coating all essential actors, automotive manufacturers, coatings manufacturers and installation constructors are
globally active, large-scale enterprises. Due to the complexity of the technical
innovation system, intensive collaboration is vital for development of new
technologies. Increasingly, coating is performed as a service within the scope
of system partnerships, where the coating manufacturer provides material and
personnel for the entire surface treatment and receives a previously agreed
price for each finished coated car body (CPU scheme = costs per unit).

(4) Innovation process: Water-based low-solvent dipping coatings have been
state-of-the-art for some time as primers. In the mid 1980s the development of
water-based fillers and topcoats was advanced by VW. These have become
established extensively in Germany. Solvent-free powder clear coatings have
been used by DaimlerChrysler (as powder slurry suspended in water) and by
BMW since 1997.

(5) Innovation drivers: Emission-related regulations, such as the Federal Immis-
sion Control Act and the EU VOC Directive were important driving factors.

The innovations described are, however, beyond fulfilling limit values. Reducing waste costs, technology leadership and the image gain related to this are the objectives of the pioneering enterprises, which implement environmentally compatible innovations in this area despite high investments and risks.

(6) Direction of innovation (from the standpoint of market actors – 6a): Automotive manufacturers substantiate the reduction of solvent emissions by water-based and powdered coatings on the basis of emission statistics. However, the automotive manufacturers state the improvement of the surface quality with the simultaneous improved cost-effectiveness of procedures as the decisive criterion for successful innovation.

Direction of innovation (from the standpoint of the SubChem research group – 6b): Further operating figures for integrated evaluation (e.g. technology comparisons to determine the "best available technique" as per IPPC directive) are less easily accessible. The higher drying energy has to be taken into account when evaluating water-based coatings. In the case of dry-applied power coatings, however, no water is used, which facilitates higher energy efficiency. Waste is also minimised, as excess coating powder is fed back into the process.

(7) Conclusion with regard to stock of hypotheses and model: The globally active enterprises share a common goal and guiding principle (technology leadership for high-quality, economical and environmentally compatible processes, Hypothesis 10). As a result of the proximity to the end consumer, automotive manufacturers have a vested interest in a positive brand image, which also includes environmentally aware production processes (Hypothesis 8). The CPU partnership makes resource-saving material usage attractive also for coating manufacturers (Hypothesis 5).

Case study "UV drying printing inks in packaging printing"

(1) Technical advantage/function: The main purpose of a colour impression on a package is to transport information. A fast processing (drying on fast running machines) and a good adhesion have to be guaranteed. Furthermore the taste and the smell of the packed commodity should not be affected. Curing and drying of the ink can be carried out by evaporation of the contained solvent (heat and air circulation) or by chemical reaction which is initiated by ultraviolet radiation (polymerisation of acrylates with photo initiators) which is particularly fast.

(2) Environment and health-related problems: About 30% of solvent emissions by printing industry are related to packaging printing - in Germany this represents an amount of approximately 16 000 t/a. Packaging printing processes are not only solvent based, but also water based with products still containing up to 20% of solvents. These printing processes represent a main source of VOC-emissions and thus contribute to the formation of ground level ozone

(summer smog). Solvent related risks are furthermore an issue in occupational health and safety. UV-drying systems do not give rise to these kinds of solvent related risks, but may under certain conditions cause skin irritation for workers.

(3) Market and actors: A variety of companies contribute to the production of a printed packaging foil: Manufacturers (packaging foil, acrylates, photo initiators and printing inks), engine construction (printing machines and additional equipment like UV lamps), repro services, packaging print shops and the manufactures of the commodities. 1997 about 18 000 tons of flexoprinting inks and 35 000 tons of gravure printing inks (55% of them solvent based) have been applied in Germany. Since their bringing onto the market UV-drying printing systems have reached a market share of only 3% of the overall flexoprinting market. Solely for gravure printing on sheet metal and for finishing varnishes a relevant market share of 21% resp. 59% was reached.

(4) Innovation process: The first patent for UV-drying paint was applied in 1946. 1971 UV-systems were launched for offset printing and 1992 for packaging flexoprinting. 2002 about 1250 equipments for UV-drying systems were installed in 600 print shops. However, in the meantime, most of the big UV-flexoprinting machines are disused or converted to solvent based systems. The reason for this is that UV-systems depend much more on well skilled personnel in order to work free of trouble. The former quality advantage over the solvent based systems has been outweighted in the meantime. The determining factor for high quality printing is the adaptation of the colour system and the frequently changing packaging foils with varying properties, which is more difficult with UV-systems. Incommoding odour can only be avoided by high efforts.

(5) Innovation drivers: The innovation of UV-curing printing inks was mainly promoted by the innovation willingness of a few print shops and their consulting engineers and of printing ink manufacturers. At the beginning of the nineties incentive for the innovation in packaging printing was the limitation of solvent emissions by the Technical Instructions on Air Quality (German TA-Luft). Print shops had to decide on whether to invest in exhaust air treatment or in conversion to low solvent systems. Companies did not succeed in ensuring adequate adhesion of water based printing inks on packaging foils and many of them went for exhaust air treatment. UV-systems were established merely in the relatively slow flexoprinting of labels on certain foils. Most of the smaller print shops were not confronted with the higher requirements of the TA-Luft as long as they did not expand and in that way have not been forced to convert.

(6) Direction of innovation (from the standpoint of market actors – 6a): Some companies made bad experiences with skin irritation caused by acrylates and photo initiators in the launched UV-drying printing inks (warning by employers´ liability insurance and trade union 1981) and the formation of benzene in cationic processes.

Direction of innovation (from the standpoint of the SubChem research group – 6b): Solvent free packaging printing processes may contribute to reduction of VOC-emissions and of solvent related risks at the workplace. A similar reduction level can not be expected with improved contained conditions for solvent based systems. Energy consuming air circulation and afterburning can be avoided with UV-systems. However, energy is needed for supply of inert gas and cooling of the printing machines.

(7) Conclusion with regard to stock of hypotheses and model: The penetration of the market of an innovation with technical advantages can be hindered, if a complex network of multiple actors with own interests leads to system inertia (hypotheses 10). Difficulties in adaptation on frequently changing materials and the risks to increase the share of rejects have been emphasised. Banner headlines on worse effects of acrylates and photo initiators from the beginning of the eighties still impact the product image, since the available data and assessment methods were not sufficient to obtain transparent and comprehensible risk assessment (hypotheses 4). The regulatory push from the beginning of the nineties did not promote the UV-systems, because with end-of-pipe solutions the requirements were met with less reorganisation of processes.

Case study "low-chromate cement"

(1) Technical advantage/function: Cement is an anorganic, finely ground substance (CaSiAlFe oxides), which when water is added produces cement paste. Cement is contained in numerous construction products (concrete, mortar, tile paste etc.). Chrome or chromates are "naturally" contained in the cement raw materials (e.g. limestone, clay) and do not perform any technical function in cement. Low-chromate cement is produced by adding a reducing agent (generally iron(II)-sulphate).

(2) Environment and health-related problems: Water-soluble chrome(VI) compounds in the wet cement or mortar have a highly sensitising effect and are up to 90% the cause of allergic cement dermatitis (cement eczema, "bricklayer's itch"). The high alkalinity (pH = 13) of cement aids the development of this contact eczema. "Bricklayer's itch" is one of the most frequent professional diseases in the construction industry.

(3) Market and actors: In Germany approximately 32 m. tons of cement are manufactured, of which about 16% are processed manually, 31% are used in factory manufacture of concrete parts and 53% are used in the manufacture of ready-mixed concrete. Around 5.5 m. tons are imported. The cement industry operates in a Fordist structured market for mass products (lengthy product cycles, dominated by price competition) with trends towards monopolies. Raw materials suppliers and downstream businesses (e.g. cement transportation enterprises) are also being controlled increasingly by the cement manufacturers.

(4) <u>Innovation process</u>: Although the cement manufacturers could have manufactured low-chromate cement by making a slight change in the production process and this change was already made in Scandinavia in the 1980s, German manufacturers only took action in the late 1990s. In 1999 the cement manufacturers, employers' liability insurance, trade unions, industrial associations and inspectorates agreed on branch specific regulations on "low-chromate[112] cements and products". The new directive 2003/53/EC relating to the chromate-free production of cement came into force on 17.7.2003[113].

(5) The high costs (approx. € 30 m. per annum) of the construction employers' liability insurance, caused by incidences of "bricklayer's itch" are important <u>innovation drivers</u>. Since the mid 1970s the construction employers' liability insurance, has endeavoured in consultations with the Association of German Cement Works to obtain a reduction of the chromate content in cement. Further initiatives such as e.g. GISBAU[114] and also regulations have stepped up the pressure on cement manufacturers.

(6) <u>Direction of innovation (from the standpoint of market actors – 6a)</u>: The causality between incidences of "bricklayer's itch" and the chromate content of cement has been suspected since the 1950s and was established conclusively in the early 1970s. However, it was contested by the cement industry. The labelling "low-chromate as per TRGS 613" meant that both commercial users and DIY users were able to clearly select low-chromate cement.

<u>Direction of innovation (from the standpoint of the SubChem research group – 6b)</u>: The use of low-chromate cement may result in a significant reduction of sensitisations to chromate or new incidences of "bricklayer's itch". The product security achieved, however, depends on whether chromates are reduced as an integrated part of the process or downstream (limited storage stability of reducing agent in finished cement).

(7) <u>Conclusion with regard to stock of hypotheses and model</u>: The obstructive position of the cement industry, which has been documented on several occasions, is striking. In a Fordist structured market the cement industry has little reason to push forward substitution ("Co-operation as a blockade" – Hypothesis 9).

Case study "dichloromethane-free stripping agents"

(1) <u>Technical advantage/function</u>: Numerous methods are available for removing old paint coatings from doors, windows and walls ("paint stripping"). The fastest of these methods is to use chemical stripping agents containing di-

---

[112] In accordance with TRGS 613 dated 1993

[113] Throughout Europe as of 17.01.2005 only cements with a maximum chrome VI ratio of 2 p.p.m. may be used and marketed for activities where skin contact is possible.

[114] Hazardous substance information system of the construction workers' statutory accident insurance

chloromethane (DCM). 75% of the paint stripping in handcraft and for DIY was 1998 carried out with DCM. Chemical alternatives for DCM are available (e.g. esters based on adipic acid), but their market share increases rather slowly.

(2) Environment and health-related problems: DCM is toxic for the central nervous system, for the liver and the kidneys (MAK-value 350 mg/m^3), and it is absorbable via skin. Furthermore DCM is a suspected carcinogen (classification carc. cat 3). DCM is low volatile and the vapour is heavier than air, thus high concentrations may occur at ground level during application. Activated-carbon-filters are ineffective and normal glove materials are penetrated within a few minutes. The required breathing equipment is, however, hardly used by craftsmen, which results in several deaths every year.

(3) Market and actors: DCM is produced in Europe at 10 sites, in general as a by-product of chloroform (raw product for HCFCs and teflon). 150.000 t/a are purchased in Europe (thereof 30.000 t/a for paint stripping, thereof 50% in building crafts and for DIY). Because of the coupled production the manufactures are forced to bring the substance to the market, which results in elasticity of prices. The market share of the costlier dibasic esters has reached so far ca. 1.000 t/a. DCM containing products for handcraft cost about 3 €/l. Comparable products with dibasic esters cost 5 €/l. Taking into account the higher effort for safe application and the lower efficiency of DCM-containing products (15 €/m^2 facade compared to 9 €/m^2 with dibasic esters) they are less economic.

(4) Innovation process: Small innovative companies started in the seventies to develop alternative stripping agents. However, up to now these products represent only a niche market. Even though the DCM market declined between 1986 and 2000 in Germany (about 80 %), substitution of DCM was limited to aerosols and adhesives. Also improved emission reduction in industrial plants lowered the consumption of DCM.

(5) Innovation drivers: Alternative paint strippers have been developed as a consequence of the „chlorine debate". However, neither the detailed requirements related to occupational health and safety (TRGS 612 and 212) nor the lower efficiency (cf. (3)) clearly supported the penetration of the market with the alternative paint stripping systems. One barrier may be that the users have to change their work and purchasing procedures to apply the alternatives successfully. By end of the nineties the construction employers´ liability insurance started, based on the public media, an information campaign about the risks of DCM-containing paint strippers.

(6) Direction of Innovation (from the standpoint of market actors – 6a): The risk determining properties of DCM are the relatively high vapour pressure connected with a high density of the vapour and a medium toxicity. The health and safety provisions for the use of DCM are sufficient to inform about risks

and less risky alternatives, but due to the often deficient safety data sheets the users are rather poorly informed.

Direction of innovation (from the standpoint of the SubChem research group – 6b): Some of the available alternatives, particularly the dibasic esters, are of lower risk mainly due to their lower vapour pressure. In contrast high volatile DCM-substitutes like methanol, can not be regarded as a suitable alternative for environmental and health purposes.

(7) Conclusion with regard to stock of hypotheses and model: DCM in handcraft uses represents a case, where neither fatal accidents nor hazard classification (R40) nor detailed and elaborated occupational safety instructions result in a considerable demand for safe alternatives. Reasons for this: The lack of broad enforcement of the occupational health and safety requirements and of the prohibition of self-service in DIY-stores (hypotheses 6). Producers and trade do not provide with sufficient information for their clients, to sensitise the users for safe handling (hypotheses 12). DCM-products are regarded as all-purpose paint strippers with low prices and the rather conservative behaving craftsmen do not see any reason to demand for alternatives. Handcraft services are not considered as applications of chemicals close to the private consumer (hypotheses 8). Thus, restrictions on use or a complete ban seems to be the only way to bring about a shift away from hazardous, yet technically highly efficient substances, to more safe-to-apply products.

Case study "Solvent-free decorative paints made of raw materials"

(1) Technical advantage/function: Varnishes are used to protect coated materials (e.g. wood, metals) in internal and external applications. However, they also have an important aesthetic function and effect on the living environment.

(2) Environment and health-related problems: "Bio varnishes", i.e. varnishes based on natural, renewable raw materials, were developed as *"close-to-nature"* alternatives (substitutes) i.a. as a reaction to the so-called 'German wood preservative scandal' and indoor pollution due to chemical solvents. Nevertheless, they have until now had a relatively high content of volatile bio-organic solvents, which may cause irritations, allergic and neurotoxic reactions, and contribute to the formation of tropospheric ozone.

(3) Market and actors: The market for bio varnishes is rather a niche market (accounting for approximately 2% of the entire paint and lacquer market). Decisions to purchase bio varnishes are comparatively informed and conscious (at least this is still the case), similar to the decision to purchase organic foods (i.e. this is an ethical market). The manufacturers try to win over their customers' confidence by means of brand strategies and transparency (full declaration). Germany is the 'lead market' in this field.

(4) Innovation process: Unlike the almost perfect ecological performance of bio varnishes, the toxicological aspects of bio-organic solvents did represent a weakness. With support from public funds, one manufacturer has now successfully developed water-based bio varnishes. The interesting aspect is, on the one hand, the ability to be innovative of small and medium-sized enterprises, which in this case is largely characterised by a strong entrepreneurial figure; on the other hand it is also the way of dealing with conflicts of goals (renewable raw materials versus reduction of summer smog and skin irritations).

(5) Innovation drivers: Besides the typical motivations in an "ethical market", the competition with water-based synthetic varnishes (which in Germany feature the "Blue Angel" label) also spurs on innovation. In addition, the announcement of an EU directive has also contributed to limiting the solvent content of decorative paints (Decopaint Directive 2004/42/EC).

(6) Direction of innovation: The use of renewable raw materials and, closely linked with this, the compostability of production waste are trend-setting. The absence of solvents was not "traded in" by using other problematic substances (such as e.g. chemical emulsifiers, drying agents or biocides). In professional applications the idea of 'water on wood' and the comparatively long drying times are, however, considered with some reservations – despite the fact that the technical performance is in no way behind that of conventional products (EN certification).

(7) Conclusion with regard to stock of hypotheses and model: Quality competition and customer proximity open up scope for impact, especially if the quality is also documented in consumer testing journals. Small companies, led by a strong entrepreneurial figure, guided by a worked out guiding principle (soft chemistry) and with the backing of public funds, are able to undertake extensive innovation steps (Hypotheses 8 and 11).

Case study "environmentally sound textile auxiliaries"

(1) Technical advantage/function: Textile auxiliaries are applied as auxiliaries in textile processing and finishing (e.g. size, cleaning agents) or as finishing agents remaining on the textile fibre (e.g. fabric softener, optical brightener). Alkylphenol ethoxilates (APEO) represent a group of environmentally dangerous process auxiliaries (tensides) in textile finishing. These substances reduce (the) surface tension between the textile fibre, water and non water soluble substances and thus support the exchange of substances between the fibre and the liquor (liquid?). Process auxiliaries are normally discharged with waste water. Same applies to product auxiliaries which tend to ensure a certain appearance (e.g. a soft grip) of the product until it is sold. These substances are washed out by the textile customer.

(2) Environment and health-related problems: APEOs, discharged into sewage treatment plants or the environment, are degraded to alkylphenols. These substances are very toxic to algae and can have endocrine disruptive effects on water organisms. Alkalphenols are only slowly degradable and have the potential to bioaccumulate in water organisms. The existing chemicals work in the EU identified a need to reduce the risks related to the use of nonylphenol ethoxilates. The German association of the auxiliaries producers (Textilhilfsmittel-, Lederhilfsmittel-, Gerbstoff- und Waschrohstoff-Industrie: TEGEWA) introduced a system of classification of textile processing chemicals according to their relevance for (to?) waters (Abwasserrelevanzstufen: ARS) to reduce environmental risks related to the use of textile auxiliaries in general. This system classifies textile auxiliaries into the categories "low relevance for (to?) waters (ARS I)", "relevance for (to) waters (ARS II)" and "high relevance for (to) waters (ARSIII)" based on criteria for human and ecological toxicology.

(3) Market and actors: About 4,600 textile auxiliaries (containing 400-600 different substances) are marketed in Germany for various processes of textile finishing. A typical textile finisher applies often more than 100 different products. About 130,000 tons of textile auxiliaries have been sold in Germany in 2002 by ca. 30 producers and used by ca. 500 textile finishers.

(4) Innovation process: Waste water issues have been discussed in Germany for more than 100 years (colouring agents, oxygene depletion, odours). In the nineteen-eighties also certain substances (e.g. environmentally dangerous properties of APEOs) have been subject to the public debate. The introduction of the TEGEWA-system in 1997 was the first systematic assessment approach that addresses self responsibility of the economic actors. The German association of textile finishers (Textilveredelnde Industrie: TVI) recommended already 1997 to use only those products, which can be classified as ARSI (low relevance for (to?) waters). The first monitoring report regarding the implementation was released in March 2001, the second in March 2003. This document reports the increase of products with low relevance for (to?) waters between 1997 and 2002 from 63% to 80% and the decrease of products with high relevance for waters (ARSIII) from 18% to 4%.

(5) Innovation drivers: The auxiliary producers implemented the system of classification to facilitate the communication between textile finishers and water authorities and to avoid additional regulatory requirements for textile finishing. Furthermore they created a system, which enables them to characterise environmental properties of a product without the necessity to disclose their recipes.

(6a) Direction of innovation (from the standpoint of market actors): The ARS-system was implemented by every auxiliary producer and was known at least by 90% of the textile finishers. The prevention of products with high relevance for waters (ARSIII) is a common interest of all actors in the supply chain.

(6b) Direction of innovation (from the standpoint of the SubChem research group): The ARS-system supported perception of responsibility for assessment on the producers' side and the demand for environmentally friendly products on the users' side. In this case marketing and systematic product assessment based on environmental criteria complement one another and thus make an impact. Concerning the method for environmental degradation there is a need to adjust to the EU-TGD (ready biodegradability of separate substances).

(7) Conclusion with regard to stock of hypotheses and model: Two important actors of the supply chain (auxiliary producers and textile finishers) have been involved in the ARS-system. The clothing industry and the commercial enterprises (bulk good) did not participate. Although the latter supply directly to the customers, assessment systems concerning process related emissions (ARS or ÖKOTEX 1000) have not yet been of interest (hypotheses 8). The ARS-system and its environmental requirements for waste water motivated the commercial actors in the supply chain to apply to (to strive for?) more transparency and for an environmentally friendly (orientation in the choice of products (hypotheses 7). In addition, (at the) auxiliary producers research activities have been activated. (hypotheses 2).

4.3   Hypotheses as a means of detection and a form of result

Working on case studies with the objective of comprehending correlations and implementing in hypotheses (hypothesis generation) is a research strategy that is entirely appropriate for the present status of innovation research. As such an approach to research necessarily requires a rather qualitative procedure and is based only on a few case studies, from the outset there was no claim to verification in respect of quantitative significance.

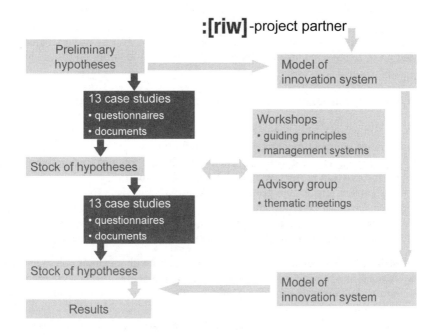

Figure 17. Operation process of model development

The hypotheses drawn up right at the start of the project had the function of directing attention to certain questions and correlations when processing the case studies. It could then be stated as a result of this focus that the applicable hypotheses in each case were confirmed or were rather refuted by the case study. Frequently, however, it transpired that some of the hypotheses were rather irrelevant for this specific case.

As we considered an improved and extended set of hypotheses right from the start as a desirable project outcome, we did however not stop with these findings. Rather we endeavoured to specify the individual hypotheses and also to develop new hypotheses beyond these. Our set of hypotheses was thus not rigid, i.e. we did not develop hypotheses in order to falsify or verify them using a sufficiently representative number of case studies. Rather, the hypotheses were further developed (and in some cases also brought to a culmination) both when analysing the case studies and also on the basis of our theoretical work using the model in such a way that they express the existing conditions adequately at least for some cases. In specific terms this means that a hypothesis can be "illustrated" quite well by a case study and that the same hypothesis may, however, also be refuted or simply be irrelevant in other cases. The hypotheses are also based on a very different degree of abstraction. They refer for example to both the general framework conditions and also to the restricting and promoting factors within the particular supply chains, always with regard to the various actors and their possibilities for action.

For the more case-specific correlations the revised hypotheses represent the central statement of results. The hypotheses with further potential for generalisation were also still included in the model formation. In this way extrapolation of our set of hypotheses also formed an important input for work on the model.

4.3.1   Hypotheses – factors promoting and restricting substitution

A set comprising 15 hypotheses was compiled in the course of the project. These hypotheses are presented and explained in the following:

Hypothesis (1): Current chemicals legislation creates competitive advantages for existing materials[115], which impede innovation

The current registration procedure for new substances in Europe (high information requirements even for small quantities), together with the absence of binding test requirements for existing substances, represents a hindrance for substance innovation in the area of high-volume, "established" industrial chemicals.

- 90% of all new registrations refer to substances with a market volume < 10 t/a
- Around one third of all new registrations is for intermediates and process regulators (for synthesis processes), and are thus rather part of a (synthesis) process innovation.
- Around 10% of all new substances registered are used for the purposes of plastics processing. A further approximately 10% of new substance registrations are in each case for (i) dyes (especially for the textile industry) and (ii) photochemicals (photographic industry and toners in the paper industry) and also (iii) cosmetics and perfumes. The majority of relevant substance innovations appear to take place in this relatively narrow market sector, while here too a relevant proportion of process innovations could be in processing (with the exception of cosmetics and perfumes).
- About 50% of new substances registered in Europe refer to imported substances from Switzerland, the USA and Japan. The number of new substances registered in the EU in this respect cannot be an indicator for the intensity of innovation in the EU Member States in the field of chemical substances.

With regard to the case studies contained in the SubChem project, the effects of the different requirements placed on new substances and existing substances can best be illustrated using the case study of plasticisers for PVC (DINCH as a substitute for DEHP).

Hypothesis (2): Under current chemicals legislation environment and health-related quality competition between chemicals is hampered by a lack of substance information

Due to the lack of legal requirements to improve the state of information concerning the chemical/physical and (eco)toxicological properties of existing substances, the information available to chemicals users remains extremely patchy

---

[115] I.e. chemical substances, which were already on the European market prior to 1981 and still account for more than 90% of the chemicals market today.

and inconsistent. A rational, environment and health-related "quality comparison" between chemicals is thus not possible from the standpoint of industrial, commercial and private chemicals users. The market actors are accordingly less sure of orientation when choosing between the available chemical alternatives.

- The products cannot be compared with regard to the environment and health-related risks, so great the level of uncertainty and the fear of making a wrong decision are. This situation encourages a conservative stance in investment and procurement decisions of chemicals users. Patchy information for "new" products are thus frequently overstressed (in many instances also for tactical reasons). This "conservative" stance is favoured even more by the fact that most users know of cases where the substitutes proved to be just as problematic or even more problematic after a few years.
- In addition, the complex status of information in combination with the relatively restricted knowledge make many commercial chemicals users susceptible to incorrect information.
- Especially concerning "substances involved in public scandals", decisions to substitute are then taken 'under pressure', in which the environment and health-related quality of substitution solutions remains to a large extent open and mere shifts of risks are probable.

In almost all cases of SubChem the factor 'lack of quality competition due to unresolved evaluation issues (either rashly negative or rashly positive evaluation)' plays a more or less major role. The following cases substantiate the hypothesis:

- mineral fibres in construction,
- anti-corrosion components in cooling lubricants,
- stripping agents,
- mould release agents with regard to aspect of biodegradability,
- substitution of hydrocarbons by esters in metal cleaning applications.

In the following cases the lack of information is clearly superseded by a scandal situation:

- phthalate plasticisers in toys for infants and babies,
- formaldehyde separators in cooling lubricants,
- substitution of CHCs by hydrocarbons in metal cleaning applications.

Hypothesis (3): The demand for "toxicological certainties" reinforces a 'lock-in'[116] for high-volume existing substances

Not only a lack in knowledge but the endeavour for safety of evaluation and decision by means of the latest (eco)toxicological certainties may result in a competitive disadvantage of new chemical solutions in comparison with "established" high-volume existing substances. The available testing methods cannot cover all eventualities. In addition, they frequently produce conflicting results, e.g. with re-

---

[116] The term 'lock-in' refers to a specific form of path dependency. A change in technology is made extremely difficult by such structural consolidation.

gard to the long-term and chronic effects of substances or the effect of substance mixtures on organisms. A strategy of data maximisation may therefore rather even lead to a deterioration of the information situation. In addition there is also a problem of resources. While the manufacturers of high-volume existing substances can easily pass on the costs for such data maximisation also via costly tests by means of the product price, the provision of information is a risky investment for the manufacturers of small-volume alternatives. A high level of toxicological certainties required by either regulations or the market can thus contribute to a 'lock-in' for the existing inventory of high-volume substances and thus, if necessary, also a reduced level of innovation, by means of which the overall possible potential for substitution is not exhausted.

One example for this volume effect is the hesitant substitution of the plasticiser DEHP in soft PVC products[117]. In this respect the attempt of the Danish environmental authorities to compensate for the competitive disadvantage of substitutes by taxing phthalates is also notable.

Textile dyes are an example of the successful introduction of new substances. Clearly positive market impulses appear to have been given due to the publication of a positive list (on the basis of the statutory registration information) by the authorities. The market for textile dyes is, however, a market with comparatively small market volumes and high market prices. For this reason, the volume advantage of existing substances plays a smaller role. In addition, the "old" dyes are now "copied" on a large scale by Asian manufacturers and are exported to Europe in high quality or not such high quality (with reference to their hazardous substance content).

Hypothesis (4): Inadequate evaluation methods and lack of qualification of commercial users prevent an "informed" product choice and increase the significance of emotional evaluation elements (e.g. preconception: "water-based = health and environmentally friendly")

In cases where comparative substance data, transparent methods and the qualification of substance users are missing for the purposes of a comparative evaluation, emotional driving forces play an especially significant role in the development of markets. This applies to "scandals" relating to substances that are not sufficiently founded on scientific facts as well as a "positive image" of certain substances or products that are not sufficiently founded on scientific facts. As scientific findings relating to substance effects come up against definite pragmatic limits (such as costs, time, manpower), ethical limits (animal testing) and limits of principles (innocuousness cannot be 'proven' positively, effects of interventions in complex and dynamic systems are not predictable), this situation cannot be tackled solely by increasing the scientific characteristics. For all actors, but especially for those enterprises relying on market success, therefore in addition to 'clarification' (i.e. the explanation of facts using scientific methods) also a qualified communica-

---

[117] The interesting aspect here is however the lead market medicinal products, which was especially important for the development and introduction of DINCH as a substitute for DEHP.

tion (comprehending and thus functioning) is important to deal with the "emotional aspects" of chemicals evaluation and risk awareness.

One example of the effect of such factors is UV inks used in printing, which were the subject of a scandal early on, the positive evaluation of 'aqueous systems' and 'bio paints', which were evaluated very positively despite their not unproblematic content of vegetable-based solvents. In addition, the comparatively speedy substitution of chlorinated solvents in cleaning processes and formaldehyde separators in cooling lubricants can only in fact be explained by the publicity-effective scandals relating to 'chlorinated chemicals' or formaldehyde.

Hypothesis (5): Chemical innovation takes place not only "in chemicals". The invention of substance applications and the "sale of advantage" instead of substances are also part of it. In addition, even extremely small market shares of products with an explicit environmental or health-related quality can indirectly become driving forces of improvement innovations in the entire market. This means that the real innovation processes in the domain of chemicals cannot be recorded adequately either solely by statistics relating to new substance registrations or solely by the classic instruments of market analysis.

Innovation in the area of chemicals is frequently reduced to the development of new substances. The registration of new substances and relevant market shares are thus interpreted as indicators of the innovation activity (or ability to be innovative) in various economic areas[118]. This indicator appears at best to be capable of indicating essential aspects of the innovation history for the area of the active ingredients (pharmaceuticals and pesticides) and industrial special chemicals. In the extensive market of industrial chemicals, on the other hand, the "invention" of new applications for 'old' substances, the reformulation of preparations or the invention of new chemicals services should play a more important role.

Application innovations, e.g. also the cross-adaptation of solutions from one branch of industry to products or processes in another branch of industry, form an important mechanism in the substitution of hazardous substances. This applies both to the demand pull as well as to the technology push. An essential reason should be in the fact that they are frequently relatively non-capital-intensive. For successful substitution of hazardous substances using this method the degree of networking between different clusters of market actors and limited legal barriers are of considerable significance for the enhancement of substance applications.

In hazardous substance substitution with a low level of innovation (e.g. substitution of hazardous substances in one preparation for other substances from the pool of existing substances and labelling of product as "environmentally compatible"), the market penetration of this alternative product does not have to be a reliable indicator for the innovation history in this domain. Even with small market shares, the availability of a substitute may act as a driving force for improvement innovations for the initial products in each case.

According to information supplied by the VCI, in the past decades around 30% of the substances on the market have been substituted for other substances from

---

[118] Fleischer, M. (2000): Regulation and Innovation in the Chemical Industry

the available pool of existing substances or for non-chemical solutions within a period of 10 years[119].

Important consequences for the formulation of the new chemicals legislation on EU level (REACH) and evaluation of its possible effects on innovation are created from these correlations:

- A too restrictive definition of the "identified use" of a substance in registration may become a constraint on innovation with regard to the development of new applications or new preparations.
- The sale of substance advantage, e.g. the sale of x m^2 of cleaned metal surface on the basis of an arbitrary combination of physical processes and the use of various 'existing substances', would not have any effect at all as a combined, organisational-institutional and substance innovation e.g. in the registration statistics of new substances.

Hypothesis (6): The user-related regulation of substances[120] alone rarely ensures sufficient demand for innovative alternatives, as it depends too much on universal execution by the authorities

The missing basic data sets for existing substances and the legal burden of evidence established with the authorities[121] for justifying market restrictions for chemical products (harmonised EU legislation) has meant that regulative interventions so far have mainly been by users (on the basis of environmental protection and occupational safety legislation, mostly in the form of legislation governing minimum standards featuring national scope for impact). Due to the large number of participants and substances it has so far rarely been a successful tactic to let the regulative impulses for a sufficient number of users become powerful enough in order to initiate a relevant demand for lower-risk alternative solutions. Both the insufficient interaction of regulation in the form of chemicals legislation (regulative push), on the one hand, and application-related employee/consumer/ environmental protection (regulative pull), on the other hand, as well as the unilateral concentration of responsibility with the state authorities are all contributing factors to the low dynamics of innovation.

---

[119] Dr. Fink, VCI; SubChem Advisory Board meeting in Frankfurt on 14th May 2003.

[120] This area, which we refer to as regulatory pull, includes in particular regulations on environmental protection, employee protection and consumer protection.

[121] This means the obligation to carry out an extensive, quantitative risk evaluation and socio-economic analysis of the alternatives being considered by the authorities with, at the same time, a legally and systematically founded lack of information.

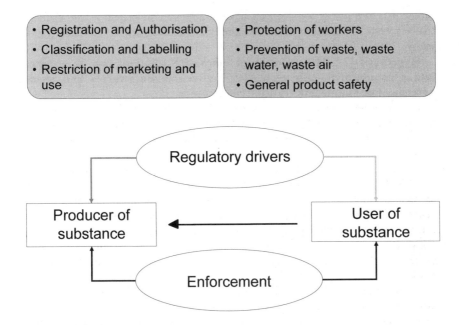

Figure 18. Regulative driving forces

The interaction of **regulative driving forces** (regulative push and pull) as well as the importance of the (limited) official execution can be substantiated well using the case studies:

In almost all cases the limits of opportunities for state intervention play a vital role with regard to the lengthy periods until regulative impulses become effective on the market. An obvious example is the use of methylene chloride in the field of handicrafts. To begin with the authorities tried to limit its use by imposing strict conditions on its use. The long path to a ban of non "biosoluble" mineral fibres in each case and the elimination of nitrosamine formers in cooling lubricants also feature among these examples. The development of criteria concerning "banned" fibre properties or "banned" ingredients of preparations took a considerable time due to the authorities' duty to provide proof.

Examples of quickly effective impulses due to user-related regulation can, however, be found in the area of chlorinated hydrocarbons: the extremely fast development in the substitution of chlorinated solvents in Germany (Tri and PER), however, can only in fact be explained by additional factors (essentially drivers in civil society). Many actors and drivers played a role here, especially the interaction of several regulative impulses, the after-effects of a Greenpeace campaign against waste disposal at sea, a trade union campaign on public health, targeted promotion of research by the state, shared interests of the chemicals industry and installation manufacturers as well as clear price signals in the area of waste disposal and for clean-up operations on abandoned hazardous sites (in this respect this is a good example of the 'multi-impulse hypothesis' from the FIU research project[122]). The prices for disposal have doubtless also played a part in the relatively speedy substitution of chlorinated paraffins (in cooling lubricants) in Germany.

Hypothesis (7): The diffusion of innovative solutions can be accelerated into a comparative risk evaluation between manufacturers and users by agreement on simple and transparent methods.

In areas where the product users and their manufacturers have agreed on simple and transparent methods and indicators for comparative evaluation of the environmental and health risks emanating from substances, the diffusion of innovative solutions (relatively small level) has been accelerated.

One example is the GisCode System for chemical construction products and the TEGEWA System for textile auxiliary materials. The VKIS list of undesired substances in cooling lubricants and the VDA list of automotive components must also be stated here, albeit with the restriction that these two lists do not offer any orientation for evaluation of the substitutes in each case, as certain criteria are missing. In addition methods for comparative evaluation, including very differing categories of effects (such as climate-toxicology-surface consumption) for the operational level, have so far not at all been put into operation for substance users.

Hypothesis (8): Environmental and health-compatible substance properties are at best additional qualities. This is particularly the case in business-to-business (B2B) markets. These qualities become most relevant for manufacturers operating with their products on demand-dominated, saturated markets with differentiated quality production (especially business-to-consumer = B2C)

The technical functionality of chemical products is the crucial product quality in B2B markets. Here especially trouble-free operating cycles and the minimisation of warranty risks play a decisive role. Environmental and health-related product qualities can only become significant for the market in places where the technical functionality is ensured and also sensitised customers or employees are available or where companies take action to secure against scandals and create an image. The additional qualities become relevant especially on demand-dominated, dynamic markets with short product cycles and differentiated quality production. However, on stable mass markets with Fordist structures even minimum price increases or quality changes may prevent innovations or at least delay them for a long time.

Examples, which back up this theory, are the success of the Ökotex 100 standard, the relative failure of the GISCODE system for paint-strippers and also the very limited success of biodegradable mould release agents in the construction industry.

Hypothesis (9): It is frequently only a small step from a net to a felt. Co-operation is a constitutive element of the ability to be innovative (beyond the enterprise). Horizontal co-operation between a few actors at one stage of the supply chain, however, may also be used to constitute blockade cartels, which are detrimental to innovation.

---

[122] Cf. the following chapter on this subject.

The actors at one stage of the supply chain are able under certain circumstances to make implicit or explicit arrangements to prevent or slow down an innovation process. It depends on the individual case whether and how this delaying tactic pays off from an economic aspect.

The conduct of the cement manufacturers in the case of chromium cements is one such case, as is the continuing resistance of mineral wool manufacturers to the introduction of the criteria "biopersistence or biosolubility" for a toxicological evaluation of mineral wools. One further example can be found in the case study examined in the course of the COIN project. In this example, the joint insistence of the three German manufacturers of titanium dioxide on the dumping of dilute acid in the North Sea in the early 1980s was ended by the unexpectedly speedy movement of one manufacturer[123].

Hypothesis (10): Co-operation networks and shared models may be an important prerequisite for innovation success.

Communication and co-operation in actors' networks supports innovations (especially in vertical communication along the supply chain). Where these factors are missing the result may be (undesired) innovation blockades because a lack of concurrence, differences in interests of the actors involved or the lack of desire to co-operate of individual important players cannot be overcome (negative network effects).

Examples of functioning co-operation are partnerships between automotive producers, coating and installation manufacturers for the development and operation of low-emission automotive series coatings and co-operation between the employers' liability insurance, IG Metall, lubricant manufacturers and metal processors in the creation of health-related standards for cooling lubricants.

Also the Quality Community on mineral wool and the industry regulation on low-chromate cement are examples of a co-operation for hazardous substance substitution that does finally function. The example of UV-drying printing inks, on the other hand, does demonstrate under which combination of negative network effects in the innovation system the diffusion of a technology can also fail (at least provisionally).

Hypothesis (11): The substance users' ability to be innovative does not solely depend on the size of the enterprise, but also just as much on the organisational structure of the enterprises as well as their position in the supply chain, for example.

The leeway and motivations for substituting hazardous substances may indeed differ between small and medium-sized enterprises (SMEs) and large-scale enterprises. However, neither of the two types of enterprises appears to be more innovative than the other per se. For example, the qualification of employees, the capital intensity of production, the availability of capital for research and development, the type of production and the competition strategy (mass or quality), the position on the supply chain as well as the corporate culture are at least just as decisive.

---

[123] Ewringmann et al, COIN 2002

Examples of innovative SMEs, also with regard to hazardous substance substitution, are to be found in the area of small ink manufacturers or manufacturers of master batches and compounds. In contrast, the speed of innovation of highly capitalised large-scale enterprises with comparatively long product cycles, such as in the automotive industry for example, may be limited. The search for substitutes for carcinogenic mineral fibres in the field of silencers and exhaust catalytic converters demonstrates, for example, that the relatively long test stages and investment cycles can hardly be curtailed, but that a potential for optimising decision-making processes does indeed exist.

Hypothesis (12): In industries with small structures with limited user know-how the business affects the willingness for innovation of the users quite considerably.

In the case of chemical products with a medium level of specialisation for use in industries with a small structure and limited user know-how, e.g. in the case of cooling lubricants, metal cleaners and strippers, the (chemicals) business plays a key role for product marketing and also for informing and advising users. There is a high level of user retention to their particular suppliers. In some cases such retention even exists between certain persons. The willingness for innovation or inertia for innovation is determined quite extensively by the industry.

As far as the SubChem cases are concerned, this theory is confirmed in the case of cooling lubricants, mould release agents and metal cleaning.

Hypothesis (13): The standardisation of products mainly has the effect of acting as a restraint on innovation, as the manufacturers tend to secure market shares of their existing products by means of standardisation. In contrast, standardisation for the procedure of taking into consideration environmental and health-related issues (management systems) may indeed have the effect of promoting innovation.

This hypothesis refers to the possibilities and limits of utilising standards and standardisation procedures to promote or prevent the substitution of hazardous substances. A sub-order was awarded within the scope of the SubChem project to this end. As a result it can be stated in a very condensed form that standardisation of products rather acts as an obstacle to innovation and is also done too frequently with the objective of securing individual competitive advantages for existing products in this "private enterprise/regulative" manner. Standardisation (and therefore also dissemination, if necessary) of (evaluation) management procedures relating to the environment and health seems to be much more promising and interesting. The ISO 9000 and ISO 14000 series are examples of this. With regard to the environmental and health-related quality of products, however, there are still major conceptual gaps between the processes of product standardisation (for example construction products) in the DIN and CEN committees and the management systems on corporate level. For this reason, the application of DIN report 108 (Guide for the inclusion of environmental aspects in product standardization and development) is not compulsory in the work of DIN.

Hypothesis (14): Comprehensive sustainability innovations can be missed by "continuous improvement" in too small stages, e.g. the substantial reduction of both resource consumption and also toxic risks.

The chance for a substantial rise in resource efficiency and an effective reduction of toxic risks can be missed if the market actors concentrate on alternatives with a lower level of innovation. Frequently there is a limitation on the side of the suppliers due to the limits of the traditional business (plastisol manufacturers do not produce underbody hard shells, manufacturers of floor coverings are bound to their machinery and can thus hardly switch over to using other materials).

Examples of such effects are the SubChem cases for the substitution of DEHP as a plasticiser in PVC by other phthalates (instead of switching over to underbody hard shells), optimisation of formulations in water-soluble cooling lubricants (instead of switching over to minimum quantity lubrication) or the use of water-based flexographic inks (instead of switching over to UV-drying printing inks).

Hypothesis (15): The substitution of hazardous substances by less hazardous substances does not always result in an optimisation of resource-efficiency and the reduction of toxic risks. For example, water-based chemical products in (at least partly) environment-open applications do not generally entail fewer risks than mineral oil-based or chemical solvents in closed systems.

The substitution of chemical solvents by water in paint and coating systems and also in cleaning processes creates complex evaluation issues, as the number of the various chemical components contained in the corresponding products generally increases. In addition, the release of persistent substances into the aquatic environment tends to be favoured by this. This therefore raises the question for assessing two suitable strategies: containment of known hazardous substances in closed systems or substitution of the mobile solvent by complex water-based systems that are scarcely evaluable from a toxicological aspect.

This hypothesis can be illustrated, for example, on cooling lubricant emulsions as well as water-based coating and cleaning agent systems.

5 Developing the model – being innovative in an innovation system

The attempt to live up to the complex innovation history using an approach based on a theoretical system already has a very longstanding tradition. However, until now innovation research was dominated by concepts of innovation systems, which either referred to regional units[124] or to certain branches and/or technology clusters[125]. The innovation systems we chose to examine are the supply chains. In most cases, these innovation systems cover several branches and also different regional levels. A simple model of this innovation system was developed early in the research process (cf. Figure 5). The aim of this model was to create a structure for the framework conditions, the participating agencies, and their options for influence. While the case studies were being processed and on the basis of the derived hypotheses, this initial model was embellished by numerous critical factors and elements (cf. Figure 19). The most important additions were the influence of public interest groups, some options for influence by the state institutions which went beyond regulation, as well as distinctions between the various types of market and competition.

---

[124] Regional or national innovation systems cf. Freeman; Lundvall 1988, Porter 1990, Nelson 1993, Edquist 1997, Braczyk 1998
[125] Cf. Breschi; Malerba 1997, Malerba 2002

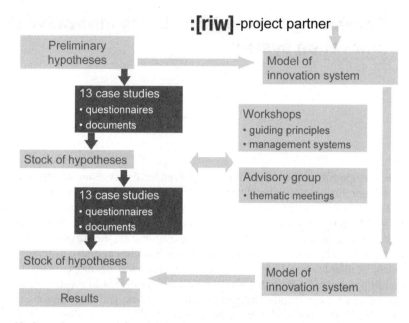

Figure 19. Operation process of model development

5.1   The framework for an innovation system

A closer examination of the case studies reveals the extreme complexity and inter-linked nature of the processes in an innovative system. Determining, which were the decisive factors that were manifest by a particular example of substitution, tended to be irresolvable in view of this complexity. A top-down analysis of the systems view of the simple model assists in orientation. In addition, some phenomena that are important for innovation processes can only be revealed from a systems view, e.g. system inertia and 'system ambience', which is frequently referred to as the 'innovation climate'. Decisively, phenomena such as 'emergence' are only discernable at a systems level. Emergence is of central importance for the comprehension of innovation processes, where 'development of a new element' is the core feature. Emergence means that a novel, unpredicted and usually complex feature is 'produced' in the system (or by the system) which no individual contributor had planned or could conceivably plan. In most cases, new elements can neither be commanded 'externally' nor can they be 'negotiated' in a discourse between the participants from their established interests. Creativity is required here, and not just the creativity of individual agencies, but rather the creative interaction of the several actors in the innovation system[126].

Our interest initially centred more on the flexibility rather than the 'creativity' of the innovation system, with the objective of better comprehending why some substitution processes are so interminably tedious, or do not progress at all, and why other substitution processes are executed comparatively quickly. The driving forces and barriers to substitution of hazardous substances were, at times, attributable to the specific case, the protagonists and agencies involved, their motives and opportunities for action. At other times, the failure was inherent to the system and the constellation of actors within the imposed framework. In the case studies, 'promoters' could be detected as well as 'opponents', who block substitution approaches categorically. For example, the insistence of the concrete industry on using chromate-contaminated cement is similar to a cartel. Similarly, the persistence of paintstrippers containing methyl chloride is bolstered by the manufacturers, with their economic interests, as well as by the professional and DIY user communities, with their interest in 'speedy and technically effective work', so that all parties have become 'immunised' against the high incidence of fatal accidents through improper use. In the majority of cases, a stalled innovation process appeared not to be actively 'impeded'. Rather it simply came to a 'standstill' and any direct 'opposition' could not be held responsible.

5.1.1   System inertia as the main barrier

The most significant barrier to innovation is 'system inertia'. Generally speaking, interested parties with comparatively significant opportunities for influence – at least within a 'non-repressive' framework – have to interact to generate 'movement in the system'. The recognition of 'system inertia' as the central barrier to innovation does tend to create an impression of powerlessness and futility. However, 'system inertia' should be recognised as the starting point for further and more differentiated examination. Detailed knowledge of essential factors, or recognition of an internal 'system logic' that could contribute to the inertia, already supply a perspective for a remedy. The ascertainment of contributory factors to the system inertia provides insights about path dependencies and technological 'lock-in'[127]. Another specific form of system inertia is constituted by investment cycles[128]. And finally it was ascertained that many innovations initially arose outside established innovation systems (and also outside large enterprises), to all intents

---

[126] The systems approach has proven to be productive not only in innovation research but also in research into technology development *(Technikgeneseforschung)*. It has superseded the previously predominant historiography, where the activities of 'brilliant' scientists and 'outstanding' engineers were recorded in succession along the more or less linear path of 'technical progress', cf. e.g. Bijker et al. 1987, Mayntz; Hughes 1988, Kowohl; Kron 1995, Weyer 1997 and 2003.

[127] Cf. David 1985 and 2000, Nelson; Winter 1982, Bijker; Hughes; Pinch 1987

[128] Erdmann 1999

and purposes 'avoiding' the corresponding system inertia. Transformation strategies in keeping with the system complexity or the system type are thus needed. Time strategies, which have recently been the subject of much discussion, with focus on the opening and closing of windows of opportunity, are important variations of this approach[129].

'System inertia' is not fixed but is being continuously modulated by various factors. From experience with the exemplified cases, two aspects require highlighting that have been discussed infrequently to date. These are the level of innovation (i.e. the degree of novelty, the size of the step, the change in trajectory, the fundamental as opposed to the incremental innovation) and the complexity of the innovative system (both with regard to technical, organisational or system innovations, and also regarding the aspects of substance, process or product innovation and/or innovation of the complete system). The importance of system inertia increases significantly both with the level of innovation and the complexity of the innovation system, i.e. the number of involved parties, to be motivated. At first glance, both phenomena are only partly interdependent. Frequently, a major innovative change presupposes the interaction of a large number of interested parties. In other cases, a small group can create innovation outside the established 'system' and shape the innovation process 'externally'. Even introductions of aqueous cleaning systems and of biodegradable concrete mould release agents, in spite of being niche products, eventually resulted in extensive innovative improvements to the products on the main market. The innovative transition can be smooth or convoluted. The substitution of CFC was reasonably facile, in which only one substance was substituted within an application system requring little adaptation. In contrast, our examination of the shift from organic solvents to water-based coatings in the automotive industry demonstrated that entire supply chains had to be reorganised and production installations converted. This dichotomy could be observed in another of the case studies covering the use of 'plasticisers in car undercoating'. The comparatively 'simple' substitution of DEHP by DINP (both high-volume substances) in the undercoating PVC Plastisol was achieved within a few months and without any extensive technical conversions. The more drastic substitution of PVC Plastisol by a plastic hard shell would incur a new underbody design and the search for new suppliers, as well as modification of the coating and assembly processes. This major step is more appropriate to a change in model (and thus linked to the investment cycle) and has only been partially implemented, even though it has existed as an alternative for more than five years.

In our experience it is not the target or the motive for innovation that contributes fundamentally to the flexibility or inertia of innovative systems, but rather the level of innovation. In accordance with our experience, hazardous substance substitution does not differ from other innovation processes that are driven by more economic or technical factors. Even comparatively simple substance-related innovations are not undertaken just to reduce risk. It is usually a technical matter of formula optimisation or a substitution of raw materials that has become too expen-

---

[129] Cf. e.g. the texts in the theme 'Transition strategies' in: Ökologisches Wirtschaften 2/2004

sive for various reasons. And for many of the cases we examined, the innovations for reducing substance-based risks were, upon closer inspection, designed for the purpose of securing market positions (or 'brands'). In this respect, innovative systems for the substitution of hazardous substances do not appear to differ from any other innovation system.

5.1.2   Competition as the main driving force

As an interim finding, it can be stated that the substitution of hazardous substances is like all innovative systems in having to combat system inertia. Innovations are desirable, but their success is compromised by system inertia. Nevertheless innovations continue to occur, implying a strong 'driving force' behind them. It is unlikely that a driving force strong enough to produce the power and the immense coordination capacity needed to set an innovative system in motion can be produced by a single institution or interested party. The driving force is as highly placed in the 'systems' level as system inertia. The strongest 'systems' driver is simply market competition. And market competition is not only motivated by success (e.g. the conquest of new markets as is usual with innovations) but also by the avoidance of losses. Stagnation is not synonymous with 'maintenance of the status quo' in a dynamic system but a reflection of deficiency and regression. In a dynamic competitive environment, companies (and the innovative systems in which they are involved) are forced to be creative and innovative just to maintain the status quo. Competition may be the driving force, but the ability to be innovative depends critically on the framework in place, the architecture of the innovative system, the specific participants and their opportunities for influence[130]. Those who cannot develop the relevant capabilities are forced out of the market.

If system inertia is the greatest barrier to innovation and if competition is the strongest driving force to overcoming it then innovation research should be devoted to the study of these two systems phenomena. This is meant not concerning such general observations, but in terms of differentiated examinations of the insightful opportunities which they offer to redesign the system. The competitive situation between companies – and now no longer only companies, but rather the entire supply chains and systems of innovation – is quite different in various mar-

---

[130] It is too presumptuous to recommend that the promotion of competition should be introduced as one of the most important measures to drive innovation. Competition is a decisive motor, but it tends only to increases the pressure and cannot always improve the capability for innovation at the same time. In some cases, 'havens' appear to be important for the development of innovation. It is no coincidence that the application areas of the military, medicine and sport are highly significant areas for technical innovation. In these areas, the cost push tends to be secondary with the result that technical effectiveness can be concentrated on fully, at least for some of the time. The concept of 'regulatory induced' initially limited 'lead markets' for environmentally friendly innovations also refers to such learning areas where new ideas and concepts can develop first of all (cf. on this latter subject e.g. Beise, Rennings 2003).

kets. In addition, much evidence indicates that competition as a whole is becoming more intensive as globalisation progresses, with market saturation and the appearance of new competitors contributing significantly.

However, it is not only the intensity of competition but also the type of markets and the quality of competition in such markets. In the absence of a scientifically based alternative, we considered two idealised market types initially:

- the 'Fordist mass market' – supply dominated, rather unsaturated, always uniform and stable, with prevailing price competition and lengthy product cycles, and a 'diversified quality market'
- the 'quality market' – demand-dominated, saturated, fragmented, dynamic and with shorter and shorter product cycles, and a strong dependence on 'brand strategies'[131].

Even on the basis of these very rough distinctions, the possibilities for action and restrictions on action for companies can be better comprehended in the case studies, by looking at the markets in which they are active. These competitive conditions constitute a broad scope for impact and 'vulnerabilities' for companies or supply chains with regard to external factors that are closely related to effects arising from their position within the supply chain.

The manufacturers of a mass product that is far from the end-user, such as cement or concrete, are under extreme price competition and slightest changes in, say, the cost structure can be decisive. Elimination of 'water-soluble chrome-VI compounds' that cause chromate-induced eczema would only increase the price of the product minimally. And even if the severe price competition did in fact permit this increased margin (which is highly probable in this example), there has been no significant innovation push. Unfortunately, this occupational disease is hardly perceived by the public and the costs incurred by this type of eczema evidently can be externalised by the cement industry (and covered by the employer's liability insurance).

However, automotive manufacturers, who practise proximity to the customer and are also known to be involved in very intense price competition, have long since undertaken action to remove certain hazardous substances from products and also from the production process, although their customers are generally unaware of or are less unconcerned by these substances[132]. End customers will not pay another cent for this and automotive manufacturers are involved in an extremely intense international competition. In order to distinguish themselves, to increase confidence in their products and to retain their customers, car companies pursue a brand strategy and are thus extremely vulnerable to 'scandals', in the press and in public. Proximity to the (end) customer, brand strategies and listings on the stock market, considerably increase the 'vulnerability' of companies to 'negative head-

---

[131] The sources of this distinction were e.g. Piore; Sabel 1984, Freeman; Perez 1988, Streek 1991, 1992, Hollingworth; Boyer 1997. We did not have the time nor the resources for a more detailed analysis and differentiation of these 'market types'.

[132] The latter is especially true for our case studies involving agents auxiliary to a process , i.e. cooling lubricants, metal cleaning agents and substances in products for the end-user.

lines' caused by their behaviour. Companies controlled by these competitive pressures cannot afford to adopt the same 'callousness' as the cement industry.

In dynamic systems that are becoming more and more complex (not least because of shorter product cycles) orientation is also becoming more and more difficult for both product suppliers and also for product purchasers. Risk communication along the supply chain is therefore also gaining in significance. Transaction costs are increasing for all parties involved. The building of confidence would pave a way for reducing complexity and transaction costs. Indeed, brand strategies rely on this principle! However, confidence has to be earned. Legitimatisation through transparency, traceability of decisions, and integration of the precautionary principle (legitimatisation by procedures) would be the ways to earn this confidence.

5.1.3   The power of scandal

One major player and systems driver has proven to be especially powerful in many cases. It is the role of publicity, the media and public interest groups and they deserve particular attention. Scandals and scandalmongering are among the most effective drivers for the substitution of hazardous substances. In some cases, public debates have influenced the state institutions as well as the more usual commercial interests.

A particularly apposite example is the campaign on the use of cleaning agents for metal surfaces (cf. chapter 4.2.1). Between 1980 and 2000, numerous public debates were concerned with chlorinated chemicals, the burning of chemical waste at sea, the negative effects of CFCs on ozone, and PER (which diffused from chemical cleaning agents into adjoining housing). A trade union campaign targeted noxious solvents and another discussion implicated VOC as the precursors of summer smog. Each issue fell to the onslaught, resulting in an impressive series of attempts at regulation at both national and EU levels. Substitution took place comparatively quickly, although the innovative system was anything but 'simple'. In particular, the area of metal cleaning is extremely complicated, varied and complex[133].

These examples demonstrate that public interest groups (i.e. trade unions, consumer protection and environmental protection organisations) can use their power to create scandals and influence companies directly, as well as affecting state regulations. The avoidance of scandal thus becomes an ever more important motive for companies and regulatory institutions, in accordance with the precaution-

---

[133] With regard to the direction of innovation, doubts are also present concerning the expediency of some substitution intitiatives. The guiding principle (Leitbild) of 'aqueous and/or water-based systems' suggested a particular route that did not always withstand scientific examination. In some cases a contained application of knowingly hazardous substances, which are clearly recognised and treated as such, would be preferable to a semi-open use of aqueous systems in which the problems are frequently underestimated (e.g. the use of biocide for microbe stabilisation or non-biodegradable components).

ary principle[134]. The omnipresent creation of quality control systems within enterprises and beyond is a direct consequence of the increased involvement of the public, the media and consumers.

It can be stated as a generalisation that 'historic' or 'innovation cultural' and 'institutional' aspects play an important role (and supposedly also one that is gaining in importance) for comprehending innovation events[135]. This conclusion is reached, for example, if we examine by way of summary the stated distinctions of the **main obstacle** *system inertia* (such as path dependency, investment cycles, complexity of the innovation system and level of innovation) as well as the distinctive features of the **main driver** *competition* (such as competition or market type and especially the role of the public). Case studies doubtless have an important function for comprehending these elements and developments. On the basis of such studies the innovation system model can also be improved and developed further (cf. Figure 20).

State institutions also possess additional factors, in particular the internalisation of external costs, opportunities for the promotion of targeted research and also the 'soft' instruments of information and moderation. Their ability to 'define objectives' has already been discussed, e.g. by formulating a 'national chemicals strategy' and especially by promoting research according to guiding principles. Technology push, research and development of substance and technical possibilities (new substances, products, processes and application systems) played a rather subordinate role in our case studies. We were primarily concerned with innovations that were developed as the result of a new combination of existing substances, i.e. truly new substances did not occur. This was partly due to the criteria of case choice, covering representative problems of occupational safety and environmental protection, different types of company, enterprise size, products, processes, markets, etc.[136]. In the past three decades of the 20th century "New substances" actually played a more subordinate role in innovation within the German Chemicals industry and the branches of users. Innovations in the latter third of the 20th century were generally highly creative new combinations of existing substances[137].

---

[134] This is probably a fundamental driver for the present REACH process at EU level.

[135] Cf. the hypotheses used in the work of the research network 'Innovation culture in Germany', http://www.lrz-muenchen.de/~innovationskultur/

[136] As usual, however, they were also determined pragmatically by factors such as (enterprise) access opportunities and the project members' existing knowledge of the particular subject involved.

[137] Cf. Grupp et al 2002, Domingues-Lacasa et al 2003.

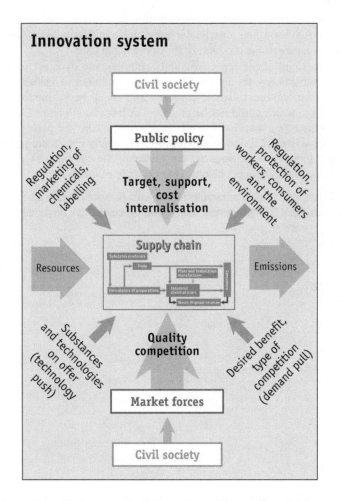

Figure 20. Interrelationship between the framework conditions, influential factors and major players: differentiated model of the innovation system

An additional important differentiation is required concerning the changing demand conditions on the markets, such as for example the shift away from mass production to a diversified quality production.

The constellations of actors in the supply chain vary between comparatively simple and linear structure with few actors, and highly networked and complex structures. Two basic types of innovative system and innovation level complexity can be deduced and to which each of the 13 case studies of the SubChem project can be assigned. As an ideal type:

- comparatively simple and straightforward systems are differentiated, in which the cause-effect relations can be assigned clearly to individual participants and their range of instruments: The cases cement, mineral fibres, concrete mould release agents are most in keeping with this type. Here it is frequently only a question of substituting a particular substance and thus maintaining the technical performance of the product for the commercial user.

- at the other end of the scale are highly complex and dynamic innovation systems, in which only the interaction of a large number of vested interests, frequently from the most diverse company and social backgrounds, can produce innovations, which could not be predicted or have been planned (emergent features). This type comprises mostly those cases where the perception of future consumer wishes plays an important role, the supply chains are part of a global network, there is no obvious system leader or the chemical products are integrated in complex process chains. One example of this is the textile chain.

If this approach to type characterisation can be successfully differentiated further, an instrument for structuring innovation processes can be developed.

6 Dealing with lack of knowledge and uncertainties – a task for risk management

Development of a model and a more profound comprehension of the complexity of an innovation process can help improve the ability of an innovation system to be innovative. It cannot provide orientation, nor ensure that the direction of an innovation is appropriate. This is the preserve of the evaluation processes, which also must cope with knowledge gaps and the unavoidable uncertainty involved in any innovation.

6.1 Direction of innovation – dealing with uncertainties and lack of knowledge

Innovation and risk are inseparably interlinked and this feeling of uncertainty is one of the most significant barriers to innovation. A successful innovation system has therefore to develop a rationale for dealing with incomplete knowledge sets and the remaining uncertainties in an appropriate way. One self-evident 'solution' for any problem of uncertainty is the acquisition of more knowledge to improve insight into the system, and this will always occur as far as is possible. However, knowledge gain does not really resolve the problem - it simply makes it more or less tangible. An improved and more accurate examination of substances with regard to their toxicological and eco-toxicological effects is incorporated into the REACH initiative for existing substances beyond a certain production volume. But data accumulation takes time and this will only be a minimum data set.

Practical limits to knowledge acquisition include the limited resources available for examining substances (i.e. funds, time, suitably qualified personnel) and ethical problems (animal testing). In addition to the pragmatic acceptance of 'not yet knowing', recognition of the fact that 'never ever knowing' plays an important role and must impose a practical limit. The multitude of substances, the exponentially greater number of possible exposure scenarios and ways of operating, and the relative inability to make forecasts about the possible interactions of very dynamic and complex systems after substance addition[138] emphasise the limits of these datasets.

---

[138] In view of this situation, the ecosystem researcher Holling refers to 'inherent unknowability', cf. Holling 1994. The question as to whether, and to what degree, the relative lack of knowledge about possible toxic and especially eco-toxic potential effects of

Innovation always means dealing with uncertainty. Also the substitution of hazardous substances cannot be based solely and not even primarily on knowledge about eco)toxicological effects. Putting the precautionary principle into practice within risk management is at least as important as improving knowledge about the effects of substances or improving the means of acquiring that knowledge.

We had originally planned to formulate an improved system for chemical assessment and evaluation, by taking all dimensions of sustainability into account and with the avoidance of simply shifting the problems. However, we in the Sub-Chem project resolved to develop more appropriate forms of dealing with deficiencies in knowledge and better ways to put the precautionary principle into practice.

These, like hazardous substance substitution, are part of risk management and thus the project focus on managerial solutions, in particular on precautionary based risk management. Two approaches turned out to be particularly promising:

1. the development and formulation of substances or chemicals, technologies and application systems determined by guiding principles (e.g. the principle of intrinsical safety), and
2. integration of the issues involved in occupational health and safety, consumer and environmental protection in quality management at the individual company level as well as along the supply chain.

6.2   Putting the precautionary principle into practice

Strategies to reduce risks emanating from hazardous substances should not be restricted to the substitution of a hazardous substance by a (hopefully) less hazardous one. A more useful approach is to consider the particular application and to consider alternative resolutions to the perceived problems. The problem could be solved either by substance substitution or by other measures, such as design modification or system process changes[139].

---

chemicals has declined in real terms, as a result of both scientific research and the systematic processing of 'negative experience' over the past decades, can hardly be answered conclusively. On the one hand, hope for a 'learning curve effect' is not unwarranted. The many negative experiences over the past 50 years may not simply be extended into the future as quasi naturally constant 'potential surprises'. On the other hand, 'new substances', 'new application systems', new technologies, such as nanotechnology, or quantitative rises in emissions of well-known substances, repeatedly present opportunities for new and surprising effects. Synergistic and multifactor effects are also especially difficult to detect, as chemicals possibly only play a role as promoters, as are effects that occur over several lengthy chains of effects or with considerable delays.

[139] One example is protection of wood by changing construction measures or various strategies for reducing unwanted germs in foodstuffs. Physical strategies of sterilisation or stringent hygienic practices in the production and processing of foodstuffs can replace the addition of preservatives.

Extending the search range to a function-related and not only a substance-related approach opens up new and interesting perspectives. Unfortunately, it does not always compensate for lack of knowledge. 'Trial-and-error strategies' are an option, but a responsible strategy requires the imposition of a reasonable boundary to the search space. Trial-and-error has its limitations where the risk is too great at any one time, i.e. in extreme cases where global or irreversible effects may ensue as the result of a single trial or error. This was and still is the case for technologies with an extremely high level of impact, such as nuclear energy, where waste stays radioactive for almost an eternity, but also in the release of self-reproducing, genetically manipulated organisms. Similarly, the massive release of persistent organic pollutants (POPs) or the irrecoverable leakage of heavy metals from ores are some of the high impact cases in which a *trial-and-error* strategy is prohibited.

But even a small-scale trial-and-error strategy has to be organised within society. As discussed in the previous section, innovations are rather improbable and disadvantaged by structural frameworks. Innovations depend upon freedom for them to be developed. At the same time safety barriers have to be formulated within which the search process can move freely. For example, possible environmental effects must be anticipated, necessitating controlled release in small increments and 'retrievability' must be ensured. Quantitative and qualitative restrictions must be imposed so that retrieval and repair options can still be effective if a trial is aborted. This approach is more successful if the persistence and spatial range of a chemical is low[140] than for persistent chemicals like CFCs and PCBs. This requires that limited 'learning spaces' or 'experimentation spaces' have to be created intentionally under technical and economic risk considerations. Small increments and a steady increase are to be preferred, accompanied by intensive monitoring of detectable consequences.

With 'reflexive modernization'[141], a more rational way of dealing with uncertainty and lack of knowledge is possible than with trial-and-error. An important approach exists in the 'characterisation' of substances or technologies. Assessments of effects essentially rely on knowledge obtained from three areas: knowledge of the triggering 'agent' (substance, technology), knowledge of the applicable target system (the exposed organism or ecosystem), and a scientifically

---

[140] The concept of 'persistence and spatial range of environmental chemicals' describes a type of product or process that is designed so that undesirable long-term effects far from their original place of use cannot occur. *Spatial Range* is the counterpart of the persistence that describes the spatial extent of a chemical's distribution in the environment. It is used in combination with the persistence as an indicator of a chemical's potential for causing widespread and long-lasting exposure. Cf. Scheringer 2002

[141] Cf. Beck 1996, p. 289ff. Beck highlights the secondary consequences of modernization, especially the dangers and risks related to unexpected consequences or through lack of knowledge. Reflexivity thus refers to the reflexivity of modernity and/or modernization to themselves. Cf. work within the scope of Sonderforschungsbereich 536 http://www.sfb536.mwn.de/ promoted by the Deutsche Forschungsgemeinschaft (German Research Foundation), especially in the area 'Political Epistemology of Uncertainty: Knowledge, Lack of Knowledge, Rationality', Lau; Böschen 2001 pp. 122-136.

substantiated impact model, which explains how accurately the agent affects the target system (e.g. biological impact measures such as carcinogenicity and reproductive toxicity, or environmental such as detriment to climate, etc.). If both the target system and also the impact model are still unknown, characterisation of the agent still remains as a starting point for implementation of the precautionary principle. Characterisation of the agent (referred to as 'hazard characterisation' in toxicology) can already provide indications of an effect spectrum, even when the target systems and the precise impact model are still unknown. So the release of chemicals demonstrating certain (bio)physical properties, which are e.g. both persistent and mobile in various environmental media, and if need be still bioaccumulative, is contrary to the demand for small increments and/or retrievability[142]. On the basis of a constantly developing understanding of various effector mechanisms at the molecular level, it is now possible to predict a spectrum of effect (QSAR)[143] from examination of chemical configurations within a limited molecular series.

This knowledge can be employed not only for the impact assessment, but also in a much more targeted way for development and design of substances and technologies. Substances and technologies do not simply exist - they are developed and designed by the contributors to the innovation system. Consequently, application of the precautionary principle must not be restricted to dealing with 'ready-made' chemicals and technologies. Environmental and health compatibility can and should be included in their development and formulation.

6.3   Substance development and technology design directed by guiding principles

Guiding principles play an important role in the development and design of substances and technologies; this is demonstrated by the results of innovation research and research of technology development *(Technikgeneseforschung)*[144]. Ideas can be developed and illustrated for how an 'ideal solution', an 'ideal substance' and an 'ideal application system' would be manifest under certain framework conditions using guiding principles. Such idealised perceptions also provide the converse - of what should be avoided under all circumstances. If it were possible to formulate (more or less explicit) successful guiding principles in respect of chemicals and application systems and their influence, either on further interpretation or the development of new guiding principles, this would open up a very interesting area of action for design options and for application of the precautionary

---

[142] This is also accounted for in the planned EU chemicals regulation. Substances that are very persistent and very bioaccumulative can be introduced subject to authorisation, even when there are no scientifically based indications of (eco)toxicological effects.

[143] QSAR = quantitative structure activity relations, cf. e.g. Jastorff, Störmann, Wölcke 2003

[144] Cf. e.g. Dierkes et al 1992 and Dierkes 1997

principle. The fundamental effectiveness of a guiding principle is hardly ever questioned, even when the empirical proof of their effect is rather rare and, in fact, very difficult[145]. It is still unclear whether, and to what extent, a guiding principle can actually be employed in a "targeted and quasi instrumental way" to influence or shape technologies[146].

Anyone wishing to influence and design innovations using a guiding principle must attempt to comprehend the effective prerequisites and modes of operation of those guiding principles that were successful. They are effective by motivating, by constituting a group identity, by coordinating and synchronising the activities of the individual participants, by reducing complexity and by structuring perception. The most important prerequisites for effectiveness also include their pictorial quality and emotionality, their orientating function as well as their reference to wishes and feasibilities likewise[147], in short, their ability to create a response in the awareness of the involved parties[148]. The pictorial quality reinforces a vivid impression and a reduction in complexity. Recognition of feasibility is important in distinguishing and rejecting "unrealistic" visions of utopia. Practical starting points should be provided. A guiding principle such as 'sustainable economic activity' could, therefore, be too complex, too abstract and too defensive. The often discussed guiding principles at the strategic level, such as resource efficiency, sufficiency and consistency are also too abstract (i.e. incorporation of the social metabolism in the natural metabolism)[149]. Guiding principles related to human needs could be more effective, such as the guiding priciple "Sustainable Building and

---

[145] Cf. Hellige 1996. For basic innovations guiding principles seem to be particularly important, in the stages prior to technological lock-in to a path and at times of upheaval prior to imminent technological changes in direction.

[146] Cf. e.g. Mambrey et al 1995; Hellige 1996; Meyer-Krahmer 1997; Kowol 1998, Blättel-Mink 1997

[147] Dierkes et al. 1992 also point out an interesting parallel between the type of innovation in areas characterised by either technology push or demand pull. Guiding principles play an important part in both types of innovation, i.e. both in the further development of technical options and also in the development and formulation of social requirements and the problems to be resolved.

[148] For the points "pictorial quality", "reduction of complexity", "structuring of perception", "motivation", "creation of a group identity", "co-ordination and synchronisation", "relation to feasibility" and "preferred instruments and typical ideal solutions", numerous interesting overlaps occur between 'guiding principles' and Thomas Kuhn's concept of paradigms (cf. Kuhn 1975).

[149] Consistency in this context should be comprehended as the qualitative and quantitative incorporation of the socio-technical metabolism in the natural metabolism (cf. Huber 2001). This can take place both by opening the substance and energy flows of the technosphere in relation to the ecosphere (e.g. by shifting to regenerative substance and energy sources and by conforming to the biological and photochemical degradability of substances) and also by a particularly effective delimitation of the technosphere vis-à-vis the ecosphere. The latter must be realised by means of "closed" applications and effective containment or by recycling that is as far-reaching and as high-quality as possible, cf. e.g. McDonough/Braungart 2002.

Living" outlined by the Enquete Commission *Protection of Humanity and the Environment* (cf. Enquete Commission 1997) or guiding principles at a medium level of implementation and operation, such as 'closed-loop material streams', bionics (nature as a model) or 'green chemistry' and 'sustainable chemistry'[150].

In the discussion of guiding principles with a view to reducing risks emanating from chemicals, a general consensus emerges that substance quality, application context and substance flow are the vital basic elements in a substance-related guiding principle. A capability for 'systems learning' must be added as an essential element, particularly in the acceptance of the intrinsic uncertainty in forecasts about developments in the possible effects of substances, necessitating a plan of action for dealing with them. This overall (and still quite abstract) guiding principle could be formulated as "Learning systems which approach a qualitative and quantitative concept of *industrial metabolism*" (meaning the metabolism between the ecosphere and technosphere).

6.4   Managing quality at the level of the supply chain[151]

The possibilities for extended risk management are not exhausted in the options arising from the use of guiding principles in the development and design of substances and technologies. The clarification of responsibilities and decision-making processes, in enterprises and along the supply chain, also plays an important role. Previously, occupational health and safety, environmental protection and consumer protection were overseen in separate corporate management systems. Current trends show a movement towards an 'integrated' management system that takes several aspects into account at the same time. However, an integrated quality management system of this type may no longer stop at corporate boundaries and, increasingly, it has to incorporate the entire supply chain. This is especially important for companies and supply chains that are especially vulnerable to public scandals and that operate close to end-customers or pursue a brand strategy. For more and more enterprises located in Germany, market opportunities are restricted to quality-based products and services because of competition from low-wage countries. The concerted development of confidence required for customer retention, and for access to the 'premium segment', does require considerable investment in quality assurance. The avoidance of scandals, liability suits or recall actions, which dent reputations, have now actually become an important driver. The integration of occupational health and safety, environmental and consumer protection into quality control at the level of the supply chain is especially promising – but is also very demanding. In view of the demands for more quality by consumers and a

---

[150] Cf. Ahrens, von Gleich, 2002, and also http://www.sustainable-chemistry.com/
[151] Cf. the contributions and findings within the scope of the relevant SubChem workshop in Chapter 6.5

greater willingness to file liability claims, the insurance companies are becoming increasingly important initiators in this area[152].

6.5 Evaluation of the workshops on extended risk management

As part of the SubChem project, two intensive workshops were conducted with the aim of learning from the experiences of scientists, companies and official institutions (cf. lists of contributions and participants in Chapters 6.5.1 and 6.5.2). The place of the workshops in the overall process is illustrated in Figure 21.

The subjects of the two workshops were:

- "From recycling management systems to sustainable chemicals – guiding principles in chemicals development and substance policy" with the sub-title "The development of solution strategies for rationally dealing with lack of knowledge with particular reference to guiding principles" and also
- "Quality and risk management – approaches at integrating environmental and health aspects in the corporate management".

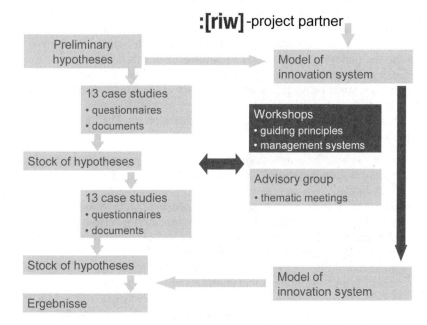

Figure 21. Workshops in interaction with the model and development of hypotheses

---

[152] Dr. Volker Kraus: Risk management from the view of the reinsurer; contribution to the workshop on Quality and Risk Management held in Hamburg in October 2003.

Both workshops were ultimately concerned with the question as to whether and to what extent the existing corporate management systems (occupational health and safety, environmental protection, product safety, quality management) can continue to be developed, so that risk management based upon guiding principles (e.g. "application-safe products" as a target perspective) can ensure that the inevitable knowledge gaps can be made practically manageable in chemicals evaluation.

6.5.1   Workshop "From recycling management systems to sustainable chemicals – models in chemicals development and substance policy"

Question

In many of the case studies, emotional drivers were identified that affect innovation processes, especially if uncertainty about effects exists through lack of knowledge. The subject of this workshop was to determine the effect of more or less explicit guiding principles on the interested parties and their contribution to sustainable development.

Decisions in research, development and innovation processes are always taken against the background of incomplete knowledge. The same generally holds true for decisions relating to the use of a particular chemical by commercial users. Extensive or even complete knowledge about possible (side) effects may thus not form the (sole) basis for substance development, substance evaluation, substance choice and substance use (application systems). The resource-related, organisational, communicative and theoretical limits in knowledge about a substance and its effects have become more than clear over the past 20 years In European evaluation and regulation processes for existing substances. The most important weaknesses include the following:

- Effect-related hazard assessment of individual substances as the most important basis for action, although the extent of knowledge about substance effects is very limited.
- The state as main actor – substance assessment by the state as the basis for risk reducing activity, mainly by means of regulatory policies. The responsibility for dealing with limited knowledge was almost exclusively that of the state and was hardly that of the other market players[153].
- Loss of confidence vis-à-vis the chemicals industry and recurring new public scandals about substances and products (harmful substance of the month).
- Lack of risk communication along the supply chain (manufacturers-trade-users), which would however be indispensable, for example, for implementing the Responsible Care scheme[154] of the chemicals industry.

---

[153] The conditions are different in the USA because of much more stringent liability laws.
[154] Cf. www.vci.de

– A lack of planning certainty in the development and application of chemicals – there is no practically manageable evaluation and orientation system.

The question is whether a guiding principle for a sustainable substance economy can make a contribution to alleviate some of these weaknesses. Ideas can be developed and illustrated using guiding principles around an 'ideal solution', an 'ideal substance' and an 'ideal application system' within a framework. Conversely, such idealisations provide a basis for what should be avoided, e.g. under precautionary aspects.

Guiding principles are thus able to provide orientation in everyday situations involving the pressure of decision-making on the basis of incomplete knowledge. Guiding principles are able to motivate innovations and provide a direction for them. They are able to coordinate and synchronise R&D activities, especially when basic innovations and a change in technological direction are made. And guiding principles are also important elements in corporate communication, both internally and beyond the corporate environment.

Guiding principles are considered as central aspects

– in providing a direction for innovation processes (change in trajectory) and
– as political, corporate and public interest groups 'control instruments' (having an influence).

There are three central issues here:

– What are guding principles? How can they be defined e.g. in relation to paradigms, technological paths, plans etc.? What effects can guiding principles have? What are the prerequisites for the effectiveness of guiding principles?
– What elements of substance-related guiding principles were effective in the past? What would be an appropriate guiding principle for a sustainable substance economy?
– If an appropriate way of dealing with the lack of knowledge is one of the central challenges facing modern societies, what contribution can guiding principles make for dealing suitably with the lack of knowledge about the effects of intervention within the context of reflexive modernization?

Effect of guiding principles

Unlike other mechanisms for orientation, such as plans, objectives, paradigms, technological paths or "dominant designs", 'effective' guiding principles have certain specific features:

– They are figurative and relate to desires and feasibilities. Values are clearly expressed in them. Guiding principles are able to co-exist with other guiding principles.
– They demonstrate a certain fuzziness, which leaves scope for flexible concretion (subtypes of guiding principles) thus also favouring the integration of practical and specific content. Guiding principles themselves in most cases do not contain any operative goals, objectives and indicators. They can thus also not

be employed as a means of direct control, but rather play a role in controlling the context.
- Guiding principles have a future-orientated perspective and refer to a desired or undesired change.

Guiding principles thus assume several functions because they:
- develop appeal, can provide orientation and contribute to the formation of identity,
- have a co-ordinating and synchronising effect on the activities of various contributors,
-, contribute to thought processes and individual motivation,
- structure perceptions,
- are a means for articulation and implementation of generalised interests in communication and negotiation processes,
- are an instrument for reducing complexity, and
- are indispensable for mobilisation and support (they enhance legitimation).

The effects of guiding principles appear to depend heavily on the extent to which they fit in with the perceptions, ideas and interests of the particular involved parties (resonance and connectivity). In addition, they must meet the contributors' capacities for action. A vital requirement for effect is also the relation of the particular corporate form, the branch of industry or the type of market and the potential to contribute in the solving of relevant problems. This also includes formulating the guiding principle at the right time (crisis relation, product cycle). A top-down introduction is also possible and is often necessary in some cases.

Contents of substance-related guiding principles

The guiding principle of a *"recycling management system"* has been developed to high degree in Germany. One explanation for this may be the overlap of a romantically excessive idea of the natural cycle with the economic idea of the money cycle. Although this guiding principle has rightly challenged the 'linear flow economy', it has favoured privatisation of waste management and caused hazardous substances to be carried over into "secondary products".

With regard to hazardous substance substitution, the guiding principle of the *"intrinsic safety"* of chemical substances, products or application systems was examined within the scope of the SubChem project. An essential approach for implementation of the guiding principle of the 'intrinsically safe substance' or the intrinsically safe application system, especially with regard to the health and ecological consequences, would be the use of 'chemicals of short ranges' (chemicals with low environmental persistence and transport distance) e.g. by:

- avoiding the environmentally open use of persistent and mobile chemicals;
- avoiding exposure to substances with delayed and at the same time irreversible, hazardous effects (e.g. carcinogenic, mutagenic, reprotoxic, allergenic);
- avoiding the open usage of mobile chemicals at the workplace (dusty or volatile substances);

– avoiding mobile components in articles of daily use (bonding of textile auxiliary materials to fibres, bonding of additives in the plastics matrix).

The guiding principle of "intrinsic safety" also contains the idea that the chemical structure of a substance already predefines its potential effects and that it can therefore be possible to determine presumably more "safe" and presumably more problematic to "hazardous" substance groups on the basis of their structure (QSAR). Further guiding principles, which were developed within the scope of the workshop, are outlined in brief in Table 2.

Tabelle 2. Further substance-related guiding principles

Guiding principle	Content:
"Sustainable chemistry on the basis of renewable raw materials"	essentially refers to the regenerative raw material basis of chemicals in relation to petrochemistry as is prevalent today.
"Ecological efficiency"	originates from the same construct as other areas of efficiency of entrepreneurial actions (costs, administration): an advantage should be created with as low as possible quantitative use of energy and raw materials.
"Ecological consistency"	refers to qualitative aspects of the metabolism between the technosphere and ecosphere. It is concerned with embedding the technical in the natural metabolism and the "matching up" of substance flows, substance qualities and conversion processes.
"Sustainable chemistry" ("Green Chemistry")	appears to require formalisation regarding products and processes in order to be capable of producing a response. The guiding principles of the energy and substance economy must be linked.
"Soft Chemistry", "proximity to nature" and "bionics"	contain the idea that "chemical-technical solutions" from nature can provide a stimulus for industrial processes and products.
"Product stewardship"	is a more management-related guiding principle, in which the substance manufacturer's assumption of responsibility and the supply chain perspective are of vital importance.

On the whole, the discussion of guiding principles has demonstrated that substance quality, application context and substance flow quantity are required basic elements in a substance-related guiding principle. Another required element is the ability for 'systems learning'. An especially important element is the uncertainty of forecasts about substance effects, accepting possible development paths as predetermined and developing appropriate procedures for dealing with them.

Dealing with knowledge limitations

The manufacture and supply of synthetic substances surpass (as do some other interventions of 'modern technologies') the needs and expectation horizons of society. In the case of "reflexive" modernization (i.e. reflecting its own bases and prerequisites), one of the central problems is dealing with "not yet knowing" and general "unknowabilities". The preferred 'alternative' at present, i.e. waiting for (toxicological) certainties before taking a decision, tends to cause stasis and systems inertia.

It is imperative to know:

– the extent of the lack of knowledge,
– the extent the possible consequences can assume and
– what options for action result from this.

For the practical implementation of this task, it is necessary to evaluate the intervention (in this case substance release) per se and in its own context (contextualisation). It is necessary to develop process types to evaluate the two conditions; knowledge and lack of knowledge (proceduralisation). A time limit for validation must be set in each case (temporalisation). The questions, methods and processes must be examined regularly (reflexive method).

The development of new substances, new products or new applications normally takes place in innovative systems. This means that the participants in the relevant supply chains can mostly be innovative only in interaction with each other or not, as the case may be. Market drivers, for the one part, affect the networks of participants and, for the other part, state initiatives, whereas interest groups in society also play an ever increasing role in both spheres of influence. This raises the question about what conditions prevail in such innovation systems which allow the development of sufficient learning capabilities in order to deal appropriately with the limits of knowledge and evaluation about substance effects.

Some system properties, which should have a favourable effect on such aspects of the ability to be innovative, include:

– transparency and clarity of the guiding principle,
– active discussion of errors and routines for improvements,
– the use of minor anomalies (mutations) as a resource and a chance for development;
– the responsibility of a particular interested party,
– the potential for formation by co-existence of a variety of solution options,
– procedures to implement product responsibility and chain management, and
– how liability law leads to internalisation of (i) evaluation costs and (ii) consequences of evaluation errors.

Conclusion

Previously, guiding principles have had a considerable influence on innovative processes and on public debates in the field of chemistry or substance economy. It can be assumed that this will also be the case in future, and that its influence will continue to increase rather than decline.

If they are incorporated in, say, management systems of quality assurance and occupational safety and environmental protection on a corporate level, guiding principles assume an important function for orientation, especially in the case of persistent knowledge inadequacies about the effects of substances.

In addition to the instruments of regulation and economic control, guiding principles are among the most interesting approaches from the group of 'informational' control instruments (3rd generation) and further examination under the theoretical aspects of innovation and control is needed and appears promising.

6.5.2 Workshop "Quality and risk management – approaches for the integration of environmental and health aspects in corporate management"

Question

Within the scope of the new EU strategy to create a chemicals policy, the state transfers the responsibility for determining, evaluating, communicating and managing the risks of chemical/toxic effects of chemicals to the industrial and commercial market actors. This responsibility relates to the manufacturing of chemicals, the processing of preparations and articles[155], use in the form of preparations and produced articles (including disposal). The workshop examined two central questions.

– How can the industry bear this responsibility (qualification, competition conditions, systematic barriers, corporate cultures)?
– How can the state support assumption of responsibility and implementation of this responsibility by creating suitable framework conditions?

When the question is raised today as to how the European production industry faces global competition, the answer is quality competition and knowledge-based product differentiation. At the same time it is noted that classic areas of the chemicals industry, such as the production of textile dyes, is practically non-existent in Europe today. This poses the questions of how do the economic actors

– organise the development of environmental and health-related product qualities (eco-design, integrated product policy),
– ensure safeguards against losses in confidence and image as well as liability claims (common in the US market), and
– apply risk reduction in a global context, in the risk management of ecotoxic effects in globally networked supply chains (like the textile chain)?

Innovation is what is new and (permanently) successful on the market. Innovation always has to overcome the inertia of existing routines (institutional structures, processes, adjustments). Innovations to improve resource efficiency and safety in chemical production have taken place in Europe and also globally in the course of the past 20 years, but have not (yet) completely replaced outdated pro-

---

[155] within chemicals legislation article means "an object composed of one or more substances or preparations which during production is given a specific shape, surface or design determining its end use function to a greater degree than its chemical composition does"

duction techniques. The formulation of chemical products and chemicals-based articles that is both efficient in resources and safe in application is less well developed by comparison and thus raises the following questions:

- How can (reliable) innovation be implemented?
- Compared with innovation processes in the automotive and IT sectors, is there a reason for the greater inertia in chemicals-related innovation?

Excessive demands on the actors in the supply chain

The general experience of public bodies and companies demonstrates that 70% of SMEs using chemicals are faced with excessive market demands with the current obligations imposed by the Dangerous Substances Order. The monitoring authorities are also faced with excessive demands to enforce its execution. Several strategies can be applied to solve the problem.

- The demands are focussed in proportion to the risk in order to concentrate the available resources on the essential risks.
- Commercial enterprises and formulators selling chemical products increasingly offer consulting services to their customers.
- Independent service-providers are responsible for determining, documenting and evaluating the risks in an external capacity.
- The qualifications of SMEs using chemicals particularly in the field of risk-related communication (both internal and external) is improved.

Transparency and procedural rules

The large number of industrial chemicals and their applications, the global nature of supply chains as well as the fundamentally limited knowledge about the direct and indirect effects of industrial chemicals on ecosystems and the human organism, all call for pragmatic restrictions for risk assessment and risk management. The risks of chemical products in each of their applications can neither be predicted exactly nor can they be reduced to zero in real-life situations.

Different scientific methods result in different risk assessments for the same substance in the same application, and one result may not necessarily be more "correct" than another result. Also the level of risk may under normal circumstances be evaluated differently by the different interest groups. In this case the relation to other risks, the sample of personal motivations and the particular context, in which the individuals are located, all play a decisive role.

This produces four important consequences for determining, evaluating and communicating risks:

- Determining pragmatic cut-off criteria, i.e. consciously accepting gaps in knowledge and uncertainty requires legitimacy. One of the essential sources for legitimacy in society is transparency, i.e. the systematic and comprehensible structuring of decision-making processes, bases for decisions and responsibility for decisions. In civilized society and in respect of dynamic global markets, the major responsibility for decision-making and risk management is left to the actors in the market.

- The inability to forecast the possible responses of the public requires a systematic and timely observation of the possible effects and correspondingly timely correction, when required (reflexiveness, ability to manage subsequently in a flexible manner).
- The processes of risk assessment, risk evaluation, decision-making, observation, re-evaluation and corrective decision-making must be structured in a comprehensible and clear way in order to create legitimacy and confidence. This includes stages where interest groups are involved (proceduralisation 1). Risk information and communication alone are not sufficient.
- The rational characterisation of risks requires iterative evaluation processes in different stages on the basis of easily available information (preliminary processes, screening). The step-by-step consolidation of information about substance properties and exposure potential should be made depending on the risk involved in order to concentrate the required resources for information procurement and evaluation where the essential risks are located (proceduralisation 2).

Procedural rules and the guarantee of transparency require institutional platforms. Such platforms can be provided by the state (REACH system or internet consultation) or, for example, can be supplied product-related by the commercial actors in the various supply chains. For the market as a whole, this raises the question as to who assumes the active leadership role within the supply chain.

Customer wishes and management systems as forces to drive innovation

The fundamental idea of the workshop was to consider the substitution of hazardous and high exposure substance applications as a problem of innovation.

Because of system inertia it is necessary for established routines to be broken and for the contributory parties to be galvanised into action. This raises the issues of what are the drivers for innovation today and under which conditions do these drivers become effective.

First consideration: Customer wishes: These raise a series of further questions for which it is less easy to find answers.

- How is product innovation determined in the 70% of the chemicals market that do not go directly to the end consumer?
- What methods can be used to determine "potential customers' wishes" in a more or less certain way and how can product quality be translated into "customers' wishes" as part of a "secure application" regarding (eco)toxic risks? The fundamental motive of "safety" alone does not appear adequate to create the requisite demand for more environmental and healthy products.
- Similar questions are raised with regard to the question as to how a demand for products and services can be created to meet "customers' wishes" in a sustainable (eco-effective) way. The quality competition for chemical-related product

safety alone does not yet resolve the problem of growth with regard to energy and substance flow rate.

Second consideration: Management systems that promote the creative potential of employees and intensify communication with the individual customers in the supply chain. This initially means interlinking more closely the sales (customer consultants), product developers, production managers and purchasing within the organisation. While the introduction of TQM and product stewardship systems can produce such effects, the innovation effects to be expected as the result of introducing corporate environmental management systems (ISO or EMAS) are rather limited. Although the induced maturing processes in the corporate management system do create scope for impact for subsequent innovations when an EMAS system is introduced, EMAS itself remains more restricted to ensuring production observing applicable legislation and can scarcely mobilise impulses by stakeholders or contribute to the development of strategic corporate objectives. The same should also hold true for the established health and safety systems at the workplace.

Quality approach

Customer orientation and initial "solution-free formulation" of customers' wishes, as an orientation for product development, appear to be promising approaches for innovations with regard to the "application safety" of chemicals-based products. However, the initiative for this is not mainly due to substance manufacturers, but rather to the chemicals users being close to the consumers. To what extent the commercial/industrial chemicals end-users (users of production auxiliary materials that are not included in the product) also transform the "latent desire" for application-safe products into effective demand behaviour, depends on other constellations of motives than those of private end-consumers. The employers' liability insurance, chambers of commerce and industry, branch associations, trade unions and management boards of large-scale companies play a key role in making "quality" and "competition" effective as drivers for innovation here too.

The example of the textile chain demonstrates (i) how much the technical/aesthetic quality of products and chemicals-related product security are interdependent and (ii) what requirements exist for a quality management system beyond the supply chain level. Only the major brands can take this initiative on a global scale. For cosmetic, medical devices and food products, as well as technical products with high safety requirements (such as aircraft), management systems beyond supply chain level have now become a matter of survival. The same holds true for products that are subject to special waste and design regulations in Europe (such as automotive vehicles and electronic equipment).

Suitable framework conditions for innovation

It is difficult to make generalisations. Distinctions are required depending on the market and constellation of actors in the supply chains, however some influential factors can be specified which increase the incidence of innovation:

- Regulatory impulses on the demand side (enforcement initiatives) create a market opportunity for innovative products and services. They are more concerned with clearly perceptible impulses (announcement effect or focal campaigns of monitoring authorities) than with enforcing regulations.
- Scandals concerning hazardous substances or product defects that have been made public in a broader context have learning effects, which go far beyond the company concerned. This means that journalistic interest in the subject of chemicals and comparative testing facilities are important drivers for innovation.
- Safeguarding against liability claims and linking insurance coverage to minimum standards of risk documentation (production and product) drive risk-reducing innovations in the area of consumer protection. The same would also be possible for employee protection if the Employer's Liability Insurance Associations were to act more like private insurers. Environmental protection against chemicals-related damage to the ecosystem or chemicals-related additional costs in the provision of drinking water or the disposal of sewage sludge could be incorporated in the specific product prices.
- If the actors in the particular supply chains (including customers, experts etc.) determine the direction of innovation in product panels or road map processes, this may have the effect of synchronising initiatives and bringing the proponents together, creating ideal conditions for targeted innovation processes. In doing so, guiding principles that are capable of producing a response (i.e. create enthusiasm) may also play an important role.
- The state provides to industry a harmonised framework for determining, evaluating and communicating risks in the European market (REACH). This also includes a form of standardisation for procedures, responsibilities and time frames in the chain, and options for state sanctions. This increases the market potential and innovation incentives for demonstrably application-safe chemical products. At the same time the basis is created for efficiency gains by employing universal IT solutions in the European market. However, the required framework also means that implementation is left to the industry and the market in many areas in order to leave enough scope for diversified solutions that are specific to the different supply chains. Simply the announcement of the REACH system has improved vertical communication in the supply chains and supported the consolidation of substance and application knowledge.

7 Recommendations for action

The SubChem project aimed to improve comprehension of innovation processes in chemicals risk assessment and from these deductions to develop recommendations for action. The recommendations are not solely related to substance and technical innovations and their framework conditions, but also encompass organisational and institutional innovations within the scope of extended risk management. From experience gathered in the course of the SubChem project, the most important recommendations for the various actors in the innovation system can be grouped according to their distinctive roles.

'Commercial institutions' include the many commercial and industrial actors in the supply chain that contribute to a system of innovation. The 'state institutions' are those to whom much more than just regulation exists as an option. And finally the consumers, the public, the media and the rest of society, who have taken power into their hands and have to learn to use it in a responsible manner.

7.1 Commercial institutions

By considering hazardous substance substitution to be part of a normal innovation process, an extensive management approach to deal with chemical substances can be established.

- Corporate management systems in the areas of environmental protection, employee protection and consumer protection should be integrated further to avoid a shift in risks and also to save costs. All areas require reliable and systematic data about substance properties and application-related exposure. Mechanisms are also required for effective dialogue along the supply chain as well as between companies and authorities. In addition, the comparative assessment of chemical products (for example, in purchasing or in product development) can only be done intelligently if all essential risk areas are included in the evaluation.
- The capacities for ambitious risk management are limited for private end-consumers and also in small and medium sized enterprises. Suppliers of chemical products for these users should therefore inaugurate the development of "intrinsically safe products" as a guiding principle. These products would be regarded as innovative because of their technical efficiency and their low risk aspect in use.

- It is also important to consider communication along the supply chains as a chance for more customer-orientated innovation. In particular, chemical traders and the formulators of preparations could expand their areas of business by adding information and consultative services (with improved customer retention).
- The acquisition of substance data and the introduction of harmonised instruments for evaluation, communication, documentation and limitation of application risks in the European market can only function at an extra-corporate level. In so doing, the following should be observed. if such solutions are in line with practice, their development cannot be left solely up to official bodies. If the costs are to be minimised, the companies have to develop standards jointly, which can however also entail the exchange of information that is relevant to a potential competitor. Thus, companies have to demonstrate courage, if efficient system solutions are to be developed at an extra-corporate level.
- The political requirement within Europe, that the industry should be responsible for the assessment of the 30,000 substances currently on the market and their applications is in line with the voluntary commitment of 'responsible care' by the chemicals industry. Implementation of this -commitment has so far failed because many users of chemical products have not complied with it. The REACH system proposed by the EU Commission would create a regulative framework for structuring responsibility and information flow along the supply chain, in a binding manner for the first time. The commercial institutions should respond to this state initiative.
- As the result of increased transparency, confidence can be re-established and the high outlay for communication within the current system can be reduced in the long term. This is especially important in areas where the identification and restriction of risks have to be stopped for pragmatic reasons or through inadequate knowledge. The aim is to reach a consensus about transparent rules of procedure. This should give more legitimacy to substance assessment and the decisions based on such assessments.

7.2   State institutions

No regulatory system can function without efficient action by the authorities. To this end intelligent inspection strategies and advisory campaigns on the part of the national and regional enforcement authorities are required which ensure that the industry complies with legislation as far as is possible (i.e. observe the agreed rules). The legally established obligation for the users of commercial chemicals to select lower-risk alternatives plays an important role in certain cases. However, substitution pressure that is exclusively regulatory does not work, as small and medium sized enterprises rarely put into practice those approaches based upon general regulations, such as the substitution principle contained in the German Ordinance on Hazardous Substances. Equally, a universal, state monitoring of regulations is unlikely to guarantee the enforcement of a general substitution principle.

This means that additional driving forces are needed, especially those that assign clear responsibilities to the actors at the individual stages of the supply chain.

However, the opportunities for state institutions to encourage innovation and its climate are not exhausted in a both innovation-friendly and risk-minimising form of regulation (creating a safety barrier). In addition to the targeted use of R&D funds directed by guiding principles (Leitbilder), the 'third-generation instruments'[156] require closer attention, such as 'information' for the various contributors and the 'moderation' of potentially competitive cooperations. State institutions are also required to curb the numerous possibilities for cost externalisation (and not just with regard to the social and health systems) in areas where it is necessary to compensate for market failures.

A number of possible strategies for successful state influence can be envisaged.

- The announcement of new regulations or initiatives for enforcement. A market opportunity is created for innovative products and services when the users of hazardous substances expect new regulations or the monitoring authorities initiate new checks. The *SubChem* case studies demonstrated that a clearly perceived initiative is frequently more effective than an attempt at the saturation monitoring of regulations.
- Efficiency gains can be boosted by integrating the various historically developed policies (chemicals, environment, employee protection, product safety). This is especially apparent in the area of chemicals safety with the first step being a change in the communication and cooperation style of the national authorities.
- The state institutions can also offer orientation (in so far as this is done on the basis of a far-reaching discourse within society). This includes the development of a national chemicals strategy, for example, or the publication and specialised validation of priority substance lists that require special attention when designing products and processes.
- The focus of the new chemicals policy should be on the practical feasibility and the transparency of systems and also must effectively transfer responsibility to the industrial sector. On the other hand, it would not be expedient to implement an extensive safety system based on scientific information and the blanket imposition of state control mechanisms.
- The ability and willingness of producing companies to co-operate is limited by the competitive situation. This is not only due to subjective interests but also concerns branch cultures and limited management resources. In a moderating and catalytic role, state institutions could initiate branch dialogue (branch agreements), new cooperations at the supply chain level as well as the development of instruments for practical communication and assessment. This can

---

[156] 'Third generation' instruments in environmental policy include 'voluntary agreements' as well as the 'soft' information instruments, such as 'best practice' examples or benchmarking, etc. The term 'third generation instruments' was created in contrast to the 'command and control' approach (first generation) and economic measures such as taxes and levies (second generation) cf. e.g. Long 1997.

reduce the need for state intervention and regulation (and the work that this inevitably involves).

- The Qualification of official bodies for consultation tasks, particularly with regard to SMEs.
- New chemicals policy should take into account that the specific costs for the industry of converting their systems must be as close as possible to the risk potential of their products. This is the only way for risk reduction to be cost effective. If the costs for the individual company are primarily derived from the quantity of substances handled and their market volume, then undesirable effects can occur. Substance manufacturers could reduce their portfolio in order to avoid costs that cannot be passed on, possibly reducing innovation in formulations, especially.
- The establishment of mechanisms to internalise external costs (liability and insurance obligations). Insurance demands do exert an influence on risk management in companies. Liability claims protection and the linking of insurance coverage to minimum standards of risk management (for production and product) are important driving forces for innovation in consumer protection. Similar principles could be applied to employee protection if employers' liability insurance were implemented more like private policies. Conversely, no internalisation mechanisms exist in the area of chemicals-related environmental costs that are caused by subliminal and long-term release of hazardous substances. Examples of this are the additional costs for the provision of drinking water, for disposal of sewage sludge or for the decontamination of buildings containing hazardous substances.
- The safe application of chemical products in the European or global market requires harmonisation of instruments and procedures to assess, communicate and to document risk-related information. This harmonisation work cannot be driven by economic factors and is one of the essential responsibilities of the state institutions.
- The development and application of a standard for "good assessment practices". The quality of risk evaluations and risk management information could become an element of competition in the global market. A verifiable and where feasible internationally recognised standard is required for this.
- State institutions can support pilot and reference enterprises by promoting innovation. Promotion of both vertical and horizontal communication (e.g. branch dialogue) and - especially for small and medium sized enterprises – the provision of basic intermediate input, e.g. free offers of information, branch benchmarks or initiate qualification programmes.
- Innovation is also driven by state-promoted R&D programmes. An emphatic reorientation of research support in accordance with guiding principles such as "intrinsic safety of products", "chemicals of low spatial range" or "sustainable chemistry" may create an impetus for innovation efforts in the industry.

7.3 Consumers and society

In recent decades, public interest groups have been instrumental in the removal of individual hazardous substances and for the implementation of the precautionary principle, in general. In doing so, the specific demand for low-pollutant products has so far only played a significant part in a few product groups (foodstuffs). Instead, the market has concentrated more upon protecting companies from the economic risks of contaminant scandals. A number of activities can contribute to consumers and public interest groups continuing to influence the discussion surrounding chemicals policy in a productive way.

- The inclusion of a sustainability perspective by all interest parties means that many previous views have had to be revised. An (eco)toxic *zero risk* or *minimal risk* is not always appropriate against the background of limited resources, the desire for innovation and when an action needs to be planned. The formulation of implementable and measurable objectives for action is a more realisable target.
- To be able to formulate objectives for this type of action requires the development of a guiding principle (Leitbild) or an agreement about an already predefined guiding principle. The "chemistry of short ranges[157]", "intrinsically safe products", "green chemistry" or a "non-toxic environment" are representative guiding principles.
- The current discussion surrounding a new European chemicals policy may result in a change of paradigm, both with regard to the level of protection and also to the means of regulation. In view of the lobbying power of the commercial institutions, the need for additional institutional capacity is evident to ensure that the public interest groups are able to contribute to the discussion adequately and participate constructively.
- One of the central requirements of a civilised society is transparency. The industrial and state authorities which create appropriate information structures do require feedback. The systematic utilisation and evaluation of offers of information is an important driver for improving the quality of information and also for developing efficient mechanisms for information access. If the public does not create any concrete demand for high-quality information then not a great deal will be achieved on the supply side.
- A market can only develop when the increased safety of chemical products is recompensed by an actual willingness to pay. Private and professional chemical users only become concerned with the health and environment-related qualities of a product (quality awareness) when they are informed effectively. Specialised journals and television news play an important role in this area.

---

[157] chemicals with low environmental persistence and transport distance

8 Prospects and prospective questions

The SubChem project has produced numerous interesting scientific perspectives which require further examination. In the area of framework conditions, questions arise relating to the quality of markets or the type of competition as the main driving forces of innovation. An increased comprehension of the markets where companies (or supply chains) operate also reveals a novel appreciation of their scope and restrictions for action as well as their vulnerability towards other 'external' influences. For example, 'market research' explicitly devoted to the types of competition would be beneficial and provide comparative observations for a more precise classification than we have been able to discern with our resources.

The recognition of the considerable influence of public pressure is particularly intriguing, where public scandals motivated both commercial enterprises and state institutions to take action in some instances. This raises the question as to what extent the power of public interest groups and the media can be influenced in a 'positive' way in order to promote the public discussion of guiding principles that could orientate the development and design of new substances and technologies.

In respect of innovation systems, the further differentiation of 'system inertia' as the main barrier to innovation is interesting. In particular, the interrelation between the complexity of innovation systems and their inertia, as well as the relation of the extent of innovative action (i.e. level of innovation) to system inertia, should be examined more closely. An important prerequisite for analysis is the formulation of a set of criteria and 'measures' for assessing the absolute level of innovation.

New opportunities for approaches aimed at the practical reduction of risks emanating from hazardous substances (beyond the scope of REACH) are two promising areas for further research and development. Specifically,

1. the prerequisites and the coverage of approaches for the development of substances and technologies based on guiding principles have to be clarified, and

2. the quality management at the supply chain level has to be further refined to incorporate, from the beginning, the concerns of environmental protection, consumer protection and occupational health and safety.

Bibliography

Ahrens, A.; Gleich, A. von: Von der Kreislaufwirtschaft zur Nachhaltigen Chemie – Leit-bilder in der Chemikalienentwicklung und Stoffpolitik, 2002 www.subchem.de/startgerman.html

Beck, U.: Wissen oder Nicht-Wissen? Zwei Perspektiven "reflexiver Modernisierung", in: Beck, U.; Giddens, A.; Lash, S.: Reflexive Modernisierung. Eine Kontroverse; Frankfurt a. M. 1996

Beise, M.; Rennings, K.: Lead Markets of Environmental Innovations: A Framework for Innovation and Environmental Economics, ZEW Discussion Paper no. 03-01, Mannheim 2003

Bijker, W.; Hughes, T. P.; Pinch, T. (ed.): The Social Construction of Technological Systems. New Directions in the Sociology and History of Technology. Cambridge, MA. (MIT Press) 1987

Blättel-Mink, B.; Renn, O. (Hrsg.): Zwischen Akteur und System. Die Organisation von Innovation, Opladen 1997

Böschen, S.: Risikogenese: Prozesse gesellschaftlicher Gefahrenwahrnehmung: FCKW, DDT, Dioxin und Ökologische Chemie, Opladen (Leske + Budrich) 2000

Braczyk, H.-J.; Cooke, P.; Heidenreich, M. (eds.): Regional Innovation Systems. London/Bristol PA (University College of London Press) 1998

Breschi, S.; Malerba, F.: Sectoral Innovation Systems: Technological Regimes, Schumpetarian Dynamics, and Spatial Boundaries, in Edquist, C. (ed.) Systems of Innovation: Technologies, Institutions and Organisations, London (Pinter Publishers) 1997

David, P. D.: Clio and the economics of QWERTY, in: American Economic Review, Papers and Proceedings 75/1985, pp. 332-337

David, P. D.: Path dependence, its critics and the quest for "historical economics", Working Paper, All Souls College, Oxford & Stanford University, June 2000 (http://www-econ.stanford.edu/faculty/workp/swp00011.pdf)

Dierkes, M.; Hoffmann, U.; Marz, L.: Leitbild und Technik – Zur Entstehung und Steuerung technischer Innovationen (edition sigma) Berlin 1992

Dierkes, M. (Hrsg): Technikgenese. Befunde aus einem Forschungsprogramm, (edition sigma) Berlin 1997

Dominguez-Lacasa, I.; Grupp, H.; Schmoch, U.: Tracing technological change over long periods in Germany in chemicals using patent statistics, in: Scientometrics 57, 2 (2003), pp. 175-195

Dosi, G.: Technological paradigms and technological trajectories, in: Research Policy 11/1982, pp. 147-162

Edquist, C. (ed.): Systems of Innovation – Technologies, Institutions and Organizations, London and Washington, (Printer Publishers) 1997

Enquete-Kommission des Deutschen Bundestages Schutz des Menschen und der Umwelt (Hrsg.): Zwischenbericht – Konzept Nachhaltigkeit, Bonn 1997

Erdmann, G.: Zeitfenster beachten. Möglichkeiten der Ökologisierung der regulären Innovationstätigkeit in: Ökologisches Wirtschaften 2/1999, 21-22

European Environment Agency: Late lessons from early warnings: the precautionary principle 1896-2000, in: Environmental Issue Report no. 22 2002, http://reports.eea.eu.int/environmental_issue_report_2001_22/en

Freeman, C.; Lundvall, B.-A.: Small countries facing the technological revolution, London and New York (Pinter Publishers) 1988

Freeman, C.; Perez, C.: Structural crises of adjustment: business cycles and investment behaviour. In: Dosi, G.; Freeman, C.; Nelson, R.; Silverberg, G.; Soete, L. (ed.): Technical Change and Economic Theory. London (Pinter Publishers) 1988 pp- 38-66

Grupp, H.; Dominguez-Lacasa, I.; Nishio-Friedrich, M. with the collaboration of Friedewald, M.; Hinze, S.; Jaeckel, G.; Schmoch, U.: Das deutsche Innovationssystem seit der Reichsgründung – Indikatoren einer nationalen Wissenschafts- und Technologiegeschichte in unterschiedlichen Regierungs- und Gebietsstrukturen. Heidelberg (Physica-Verlag) 2002

Hellige, H. D.: Technikleitbilder als Analyse-, Bewertungs- und Steuerungsinstrumente: Eine Bestandsaufnahme aus informatik- und computerhistorischer Sicht, in: H. D. Hellige (Hrsg.): Technikleitbilder auf dem Prüfstand. Leitbild-Assessment aus Sicht der Informatik- und Computergeschichte (Sigma) Berlin 1996

Hemmelskamp, J.: Umweltpolitik und technischer Fortschritt, (Physica Verlag) Heidelberg 1999

Holling, C. S.: New Science and New Investments for a Sustainable Biosphere. In: Jansson, A.; Hammer, M.; Costanza, R. (ed.): Investing in Natural Capital – The Ecological Economics Approach to Sustainability (Island Press) Washington 1994

Hollingsworth J. R.; Boyer, R. (ed.): Contemporary Capitalism: The Embeddedness of Institutions, Cambridge (Cambridge University Press) 1997

Huber, J.: Ökologische Konsistenz. Zur Erläuterung und kommunikativen Verbreitung eines umweltinnovativen Ansatzes, in: Umweltbundesamt (Hrsg.): Perspektiven für die Verankerung des Nachhaltigkeitsleitbildes in der Umweltkommunikation, UBA-Berichte 4/01 (Erich Schmidt Verlag) Berlin 2001 pp. 80–100

Jastorff, B.; Störmann, J.; Wölcke, U.: Struktur-Wirkungs-Denken in der Chemie – Eine Chance für mehr Nachhaltigkeit (Universitätsverlag Aschenbeck&Isensee), Bremen, Oldenburg 2003

Kowol, U.; Krohn, W.: Innovationsnetzwerke. Ein Modell der Technikgenese. In: Halfmann, J.; Bechmann, G.; Rammert, W. (Hrsg.): Technik und Gesellschaft. Jahrbuch 8: Theoriebausteine der Techniksoziologie. Frankfurt a.M. (Campus) 1995 pp. 77-105

Kowol, U.: Innovationsnetzwerke. Technikentwicklung zwischen Nutzungsvisionen und Verwendungspraxis, Wiesbaden 1998

Kuhn, T.: Neue Überlegungen zum Begriff des Paradigma, in: Kuhn, T.: Die Entstehung des Neuen. Studien zur Struktur der Wissenschaftsgeschichte, Frankfurt a. M. 1975

Lau, C.; Böschen, S.: Möglichkeiten und Grenzen der Wissenschaftsfolgenabschätzung. In: Beck, U.; Bonss, W. (Hrsg.): Die Modernisierung der Moderne. Frankfurt a. M. (Suhrkamp) 2001

Long, B. L.: Environmental Regulation: The Third Generation, in: The OECD OBSERVER no. 206 June/July 1997

Lundvall, B.-A.: User-Producer Relationships, National Systems of Innovation and Internationalisation. In: Bengt-Åke Lundvall (Hrsg.), National Systems of Innovation. Towards a Theory of Innovation and Interactive Learning. London (Pinter Publisher) 1992

Malerba F.: Sectoral systems of innovation and production. In: Research Policy 31(2), 2002 pp. 247-264

Mambrey, P.; Paetau, M.; Tepper, A.: Technikentwicklung durch Leitbilder. Neue Steuerungs- und Bewertungsinstrumente, Frankfurt a. M. i.a. 1995

Mayntz, R. ; Hughes, T. P. (Hrsg.): The Development of Large Technological Systems. Frankfurt a. M. (Campus) 1988

McDonough, W.; Braungart, M.: Cradle to Cradle – Remaking the Way We Make Things (North Point Press) New York 2002

Meyer-Krahmer, F.: Umweltverträgliches Wirtschaften. Neue industrielle Leitbilder, Grenzen und Konflikte, in: Blättel-Mink, B.; Renn, O. (Hrsg.): Zwischen Akteur und System. Die Organisation von Innovation, Opladen 1997

Nelson, R. R.; Winter, S. G.: An Evolutionary Theory of Economic Change, Cambridge, Mass., London 1982

Nelson, R. R. (ed.): National Innovation Systems: A comparative study. Oxford and New York (Oxford University Press) 1993

Piore, M. J., Sabel, C. F.: The Second Industrial Divide. Possibilities for Prosperity. New York (BasicBooks) 1984

Porter, M. E.: The comparative advantage of nations. London (MacMillan Press Ltd) 1990

Risikokommission (ad hoc-Kommisison 'Neuordnung der Verfahren und Strukturen zur Risikobewertung und Standardsetzung im gesundheitlichen Umweltschutz der Bundesrepublik Deutschland'): Abschlussbericht (on behalf of the Federal Ministry of Health and Social Security and the Federal Ministry for the Environment, Nature Conservation and Nuclear Safety), Salzgitter 2003

Scheringer, M.: Persistence and Spatial Range of Environmental Chemicals: New Ethical and Scientific Concepts for Risk Assessment (Wiley-VCH) Weinheim i.a. 2002

Streeck, W.: On the Institutional Conditions of Diversified Quality Production. In: Matzner, E.; Streeck, W. (ed.): Beyond Keynesianism. The Socio-Economics of Production and Full Employment. Aldershot (Edward Elgar) 1991 pp. 21-61

Streeck, W.: Social Institutions and Economic Performance, Studies in Industrial Relations in Advanced Capitalist Economies. Newbury Park (Sage) 1992

Weyer, J. et al: Technik, die Gesellschaft schafft – Soziale Netzwerke als Ort der Technikgenese. Berlin (edition sigma) 1997

Weyer, J.: Von Innovations-Netzwerken zu hybriden sozio-technischen Systemen. Neue Perspektiven der Techniksoziologie. Arbeitspapier Nr. 1 (June 2003) Universität Dortmund, Arbeitspapiere des Fachgebiets Techniksoziologie Dortmund 2003

Table of figures and tables

Glossary

- **Actors** in an innovation system means manufactures, importers and users of chemicals (actors in the supply chain, economic actors) as well as authorities, science, public interest groups and other participants outside the supply chain.
- **Innovation** is something new that is not only invented, but is also successful by actually catching on in society or on the market. It can be something fundamentally new that has never existed or it can be a refinement of an existing innovation. Old technologies can also become novelties in a different context. Innovations are also differentiated depending on the level of innovation. The improvement of an existing innovation along an established path of development is thus referred to as an incremental innovation. By contrast, a change of direction or path may produce a leap in the level of innovation. An innovation that entails a whole cascade of further innovations, e.g. the development of microcomputers, is referred to as a fundamental or basic innovation.
- **Innovation types** can be differentiated as to whether the successful new innovation is of a technical or organisational nature or signifies a change of system.
- **Direction of Innovation.** Within the context of SubChem the question is also relevant as to whether the innovation (hazardous substance substitution, hazardous substance prevention, exposure prevention) has really reduced the risks entailed by the hazardous substance or has possibly only caused those risks to shift. With regard to the practicability of risk management, the question can also be raised as to whether a specific organisational and/or institutional innovation creates increased (also practicable) personal responsibility of the commercial enterprises or whether and to what extent it is linked to a greater need for state control.
- **Hazardous substances** are chemical substances with properties that may have certain harmful effects on humans and on the environment upon contact or exposure. The risk arises from the degree of hazard presented by the substance (including any of its metabolites) as well as the nature and intensity of contact (duration, dosage, absorption route, frequency).
- **Substitution** of a hazardous substance or product signifies its replacement by less hazardous substance, product or process. In this context the scope ranges from simple substitution (i.e. exchanging substances) to risk management as a whole (i.e. prevention of hazardous substances, reduction or prevention of exposure, etc.).
- **Risk management:** The substitution of a hazardous substance contained in a product or a process by a less hazardous substance may be an appropriate way of reducing risks. Modified technical systems or methods can, however, also be considered as additional or complementary in action. Management systems for developing less hazardous substances and for the secure use of hazardous substances (closed systems, certification of users) are also part of this process. We

refer to the full package of possible or practiced measures as risk management. This can represent a component in the overall quality management process of a company or a supply chain.

- **Framework conditions, impulses and influential factors** affect the value networks externally (state, public). These include both legal and market structures and impulses.
- **Structuring or design options** describe how and with what means participants in the value network act or could act (i.e. actively or reactively) to influence the innovative process.
- As a **guiding principle (Leitbild)**, we refer to the joint vision shared by a more or less extensive group of interested parties in the public, in research and development, in state institutions or in the market, about the course of joint action. (For example, the guiding principle of a closed-loop economy was recently shown to be particularly effective example).
- **Emergence** is the unplanned and unpredictable occurrence of new system qualities and/or system entities. Examples are the increased significance of a critical audience, including public interest groups, or the emergence of new market constellations or stock exchange performance.
- **Quality** is whatever the customer considers this to be, whether explicit or not.

ABOUT THE AUTHOR

For over thirty years, Dr. Gita Arian Baack has been consulting and coaching individuals and organizations. Her interventions consistently result in positive change and individual empowerment. Her most recent work has been research on the effects of trauma inherited through generations. This has put her on a path to help *Inheritors*, as she calls the descendants of trauma, to reclaim their lives and build on their resilience. She recently founded the "Centre for Transformational Dialogue" to help individuals and communities that have inherited devastating legacies. Dr. Baack's work has helped descendants of Holocaust survivors, Indigenous people, African Americans, and Asian and other communities struggling with the aftereffects of trauma.

Dr. Baack holds a PhD from Tilburg University, The Netherlands, in association with Taos Institute, and an MA in Human Systems Intervention from Concordia University, Montreal, Canada. She is the author of *Poems of Angst and Awe,* a practitioner of Social Construction and Appreciative Inquiry, and a social activist, living and working in Ottawa, Canada.

Author photo © Hillary Baack Photography

SELECTED TITLES FROM SHE WRITES PRESS

She Writes Press is an independent publishing company founded to serve women writers everywhere. Visit us at www.shewritespress.com.

Tell Me Your Story, Tuya Pearl, $16.95, 978-1-63152-066-2. With the perspective of both client and healer, this book moves you through the stages of therapy, connecting body, mind, and spirit with inner wisdom to reclaim and enjoy your most authentic life.

The Art of Play, Carol K. Walsh, $16.95, 978-1-63152-030-3. Lifelong "non-artist" Joan Stanford shares the creative process that led her to insight and healing, and shares ways for others to do the same.

Even in Darkness, Barbara Stark-Nemon, $16.95, 978-1-63152-956-6. From privileged young German-Jewish woman to concentration camp refugee, German Jew Kläre Kohler navigates the horrors of war and—through unlikely sources——somehow findsing the strength, and hope, and love she needs to survive.

This Way Up, Patti Clark, $16.95, 978-1-63152-028-0. A story of healing for women who yearn to lead a fuller life, accompanied by a workbook designed to help readers work through personal challenges, discover new inspiration, and harness their creative power.

I Know It in My Heart, Mary E. Plouffe, PhD., $16.95, 978-1-63152-200-0. Every child will experience loss; every adult wants to know how to help. *I Know It In My Heart: Walking through Grief with a Child* uses one family's tragic loss to tell the story of childhood grief—its expression and its evolution—from ages three to fifteen.

Baffled By Love, Laurie Kahn, $16.95, 978-1-63152-226-0. A therapist offers glimpses of her own rocky history, interwoven with stories of her clients—who as children were abused by the very people they loved and trusted—creating a textured tale of the all-too-human search for the "good" kind of love.